Communications in Computer and Information Science 2131

Editorial Board Members

Joaquim Filipe, *Polytechnic Institute of Setúbal, Setúbal, Portugal*
Ashish Ghosh, *Indian Statistical Institute, Kolkata, India*
Lizhu Zhou, *Tsinghua University, Beijing, China*

Rationale

The CCIS series is devoted to the publication of proceedings of computer science conferences. Its aim is to efficiently disseminate original research results in informatics in printed and electronic form. While the focus is on publication of peer-reviewed full papers presenting mature work, inclusion of reviewed short papers reporting on work in progress is welcome, too. Besides globally relevant meetings with internationally representative program committees guaranteeing a strict peer-reviewing and paper selection process, conferences run by societies or of high regional or national relevance are also considered for publication.

Topics

The topical scope of CCIS spans the entire spectrum of informatics ranging from foundational topics in the theory of computing to information and communications science and technology and a broad variety of interdisciplinary application fields.

Information for Volume Editors and Authors

Publication in CCIS is free of charge. No royalties are paid, however, we offer registered conference participants temporary free access to the online version of the conference proceedings on SpringerLink (http://link.springer.com) by means of an http referrer from the conference website and/or a number of complimentary printed copies, as specified in the official acceptance email of the event.

CCIS proceedings can be published in time for distribution at conferences or as post-proceedings, and delivered in the form of printed books and/or electronically as USBs and/or e-content licenses for accessing proceedings at SpringerLink. Furthermore, CCIS proceedings are included in the CCIS electronic book series hosted in the SpringerLink digital library at http://link.springer.com/bookseries/7899. Conferences publishing in CCIS are allowed to use Online Conference Service (OCS) for managing the whole proceedings lifecycle (from submission and reviewing to preparing for publication) free of charge.

Publication process

The language of publication is exclusively English. Authors publishing in CCIS have to sign the Springer CCIS copyright transfer form, however, they are free to use their material published in CCIS for substantially changed, more elaborate subsequent publications elsewhere. For the preparation of the camera-ready papers/files, authors have to strictly adhere to the Springer CCIS Authors' Instructions and are strongly encouraged to use the CCIS LaTeX style files or templates.

Abstracting/Indexing

CCIS is abstracted/indexed in DBLP, Google Scholar, EI-Compendex, Mathematical Reviews, SCImago, Scopus. CCIS volumes are also submitted for the inclusion in ISI Proceedings.

How to start

To start the evaluation of your proposal for inclusion in the CCIS series, please send an e-mail to ccis@springer.com.

Gerhard-Wilhelm Weber ·
Jose Francisco Martinez Trinidad ·
Michael Sheng · Raghavendra Ramachand ·
Latika Kharb · Deepak Chahal
Editors

Information, Communication and Computing Technology

9th International Conference, ICICCT 2024
New Delhi, India, May 11, 2024
Revised Selected Papers

Editors
Gerhard-Wilhelm Weber ⓘ
Poznań University of Technology
Poznań, Poland

Michael Sheng ⓘ
Macquarie University
Sydney, NSW, Australia

Latika Kharb ⓘ
Jagan Institute of Management Studies
Rohini, Delhi, India

Jose Francisco Martinez Trinidad ⓘ
National Institute for Astrophysics Optics
and Electronics
Cholula, Mexico

Raghavendra Ramachand ⓘ
Norwegian University of Science
and Technology
Trondheim, Norway

Deepak Chahal ⓘ
Jagan Institute of Management Studies
Rohini, Delhi, India

ISSN 1865-0929   ISSN 1865-0937 (electronic)
Communications in Computer and Information Science
ISBN 978-3-031-72482-4   ISBN 978-3-031-72483-1 (eBook)
https://doi.org/10.1007/978-3-031-72483-1

© The Editor(s) (if applicable) and The Author(s), under exclusive license
to Springer Nature Switzerland AG 2025

This work is subject to copyright. All rights are solely and exclusively licensed by the Publisher, whether the whole or part of the material is concerned, specifically the rights of translation, reprinting, reuse of illustrations, recitation, broadcasting, reproduction on microfilms or in any other physical way, and transmission or information storage and retrieval, electronic adaptation, computer software, or by similar or dissimilar methodology now known or hereafter developed.
The use of general descriptive names, registered names, trademarks, service marks, etc. in this publication does not imply, even in the absence of a specific statement, that such names are exempt from the relevant protective laws and regulations and therefore free for general use.
The publisher, the authors and the editors are safe to assume that the advice and information in this book are believed to be true and accurate at the date of publication. Neither the publisher nor the authors or the editors give a warranty, expressed or implied, with respect to the material contained herein or for any errors or omissions that may have been made. The publisher remains neutral with regard to jurisdictional claims in published maps and institutional affiliations.

This Springer imprint is published by the registered company Springer Nature Switzerland AG
The registered company address is: Gewerbestrasse 11, 6330 Cham, Switzerland

If disposing of this product, please recycle the paper.

Preface

The 9th International Conference on Information, Communication and Computing Technology (ICICCT 2024) was held on May 11, 2024 in New Delhi, India. ICICCT 2024 was organized by the Department of Information Technology, Jagan Institute of Management Studies (JIMS) Rohini, New Delhi, India. The conference received 176 submissions and 49 papers were shortlisted for review, and after double-blind reviews with an average of 2 reviews per paper, 13 papers were selected for this volume. The acceptance rate was around 28%. The contributions came from diverse areas of information technology categorized into two tracks, namely (1) Intelligent Systems and (2) Pattern Recognition.

The aim of ICICCT 2024 was to provide a global platform for researchers, scientists and practitioners from both academia and industry to present their research and development activities in all the aspects of communication and network systems and computational Intelligence techniques.

We thank all the members of the Organizing Committee and the Program Committee for their hard work. We are very grateful to Gerhard-Wilhelm Weber, Faculty of Engineering Management, Poznań University of Technology, Poznań, Poland as First General Chair and José Francisco Martínez Trinidad, National Institute for Astrophysics, Optics and Electronics, San Andrés Cholula, Mexico as Second General Chair, to Raghavendra Ramachand, Department of Information Security and Communication Technology, Norwegian University of Science and Technology, Norway as Program Chair, to Michael Sheng, School of Computing, Faculty of Science and Engineering, Macquarie University, Sydney, Australia as First Keynote Speaker and Rangaraj M. Rangayyan, Department of Electrical and Computer Engineering, University of Calgary, Alberta, Canada as Second Keynote Speaker, and to Pratyay Kuila, Department of Computer Science Engineering, National Institute of Technology Sikkim, India as session chair for Track 1 and Mantosh Biswas, Department of Computer Science, University of Delhi, India as session chair for Track 2.

We thank all the Technical Program Committee members and referees for their constructive and enlightening reviews on the manuscripts. We thank Springer for publishing the proceedings in the Communications in Computer and Information Science (CCIS) series. We thank all the authors and participants for their great contributions that made this conference possible.

June 2024

Latika Kharb
Deepak Chahal

Organization

General Chairs

Gerhard-Wilhelm Weber — Poznań University of Technology, Poland
José Francisco Martínez Trinidad — National Institute of Astrophysics, Optics and Electronics, Mexico

Program Chair

Raghavendra Ramachand — Norwegian University of Science and Technology (NTNU), Norway

Keynote Speakers

Michael Sheng — Macquarie University, Australia
Rangaraj M. Rangayyan — University of Calgary, Canada

Conference Secretariat

Praveen Arora — Jagan Institute of Management Studies, India

Program Committee Chairs

Latika Kharb — Jagan Institute of Management Studies, India
Deepak Chahal — Jagan Institute of Management Studies, India

Session Chair for Track 1

Pratyay Kuila — National Institute of Technology Sikkim, India

Session Chair for Track 2

Mantosh Biswas   University of Delhi, India

Technical Program Committee

Chan Weng Howe	Universiti Teknologi Malaysia, Malaysia
R. Chithra	K.S. Rangasamy College of Technology, India
Hima Bindu Maringanti	Maharaja Sriram Chandra Bhanja Deo University, India
Sameerchand Pudaruth	University of Mauritius, Mauritius
Anoop	Kerala University of Digital Sciences, India
Abdel-Badeeh Salem	Ain Shams University, Egypt
Kalpdrum Passi	Laurentian University, Canada
Attila Fazekas	University of Debrecen, Hungary
Dian Saadillah Maylawati	UIN Sunan Gunung Djati Bandung, Indonesia
Ajune Wanis Ismail	Universiti Teknologi Malaysia, Malaysia
Purushottam Patil	Sandip University, Nashik, India
Janmenjoy Nayak	Maharaja Sriram Chandra Bhanja Deo University, India
Pang Yee Yong	Universiti Teknologi Malaysia, Malaysia
Nikhil Marriwala	Kurukshetra University, India
Christos Antonopoulos	University of the Peloponnese, Greece
Anazida Zainal	Universiti Teknologi Malaysia, Malaysia
Manoj Sahni	Pandit Deendayal Energy University, India
Nor Haizan Bt Mohamed Radzi	Universiti Teknologi Malaysia, Malaysia
Jacek Izydorczyk	Silesian University of Technology, Poland
Matt Kretchmar	Denison University, USA
Haza Nuzly Bin Abdul Hamed	Universiti Teknologi Malaysia, Malaysia
Sifat Momen	North South University, Bangladesh
Prasanta Ghosh	Syracuse University, USA
Yaw-Huei Chen	National Chiayi University, Taiwan
A. H. M. Saiful Islam	Notre Dame University Bangladesh, Bangladesh
Hsu-Yung Cheng	National Central University, Taiwan
Guanfeng Liu	Macquarie University, Australia
Anuroop Gaddam	Deakin University, Australia
Ajay Mittal	Panjab University, India
Pratyay Kuila	National Institute of Technology, Sikkim, India
Suman Mann	Maharaja Surajmal Institute, India
Jereesh A. S.	Cochin University of Science and Technology, India

Jeegar Trivedi	Maharaja Sayajirao University of Baroda, India
Vinod Pachghare	College of Engineering Pune, India
B. Surendiran	National Institute of Technology Puducherry, India
Nitin Kumar	Punjab Engineering College, India
Khalid Raza	Jamia Millia Islamia, India
J. Satheeshkumar	Bharathiar University, India
Saima Zafar	National University of Computer & Emerging Sciences, Pakistan
Arun Solanki	Gautam Buddha University, India
Pradeep Singh	National Institute of Technology Raipur, India
Suraiya Jabin	Jamia Millia Islamia, India
J. Akilandeswari	Sona College of Technology, India
Zunnun Narmawala	Nirma University, India
Upendra Kumar	Birla Institute of Technology, Patna Campus, India
Rakesh Chandra Balabantaray	International Institute of Information Technology, Bhubaneswar, India
Zareen Alamgir	National University of Computer & Emerging Sciences, Pakistan
Amina Samih	University Hassan II of Casablanca, Morocco
Wathiq Laftah Al-Yaseen	Al-Furat Al-Awsat Technical University, Iraq
Dalibor Dobrilovic	University of Novi Sad, Serbia
Mohammad Ashrafuzzaman Khan	North South University, Bangladesh
Pradeep Tomar	Gautam Buddha University, India
Arvind Selwal	Central University of Jammu, India
Surjeet Dalal	Amity University, India

Contents

Intelligent Systems

Investigating Freshmen Students' Coding Standards Challenges Using
NLP Techniques 3
   Edgar Ceh-Varela and Essa Imhmed

ParkinSense: A Tri-Sensory AI-Powered Framework for Early Detection
of Parkinson's Disease 16
   M. Akshaya and Emmanuel Joy

Analysing the Role of Post Click Factors in Generating Leads Through
Search Advertising 27
   Amit Kishore and C. Om Prakash

mCLIP: Multimodal Approach to Classify Memes 42
   M Kaab Bin Shahid, Hamid Husain, and Hira Javed

Enhancing Marine Litter Management in the Gulf of Aqaba Through AI 56
   Mohammad Wahsha, Heider Wahsheh, and Tariq Al-Najjar

Solving the Rush Hour Puzzle Problem by Different Heuristics 68
   Yannick Schmid, Rolf Dornberger, and Thomas Hanne

Advancing Precision Agriculture: Machine Learning-Based Crop
Recommendation for Optimal Yield 80
   Mohamed Bouni, Badr Hssina, Khadija Douzi, and Samira Douzi

Pattern Recognition

Low-Cost Biodegradable Composite-Shaped Split Ring Resonator (SRR)
Based Sensor for Breast Cancer Assessment 97
   Madan Kumar Sharma, Degala Satyanarayana,
   and Abdullah Said Alkalbani

XR-Menu: Food Ordering System in Extended Reality Application 109
   Ajune Wanis Ismail and Fazliaty Edora Fadzli

Predicting Crime Hot Spots Using Machine Learning Algorithms: Cities
in USA and South Africa 123
   Dane Brown and Anil Abraham

Experience with the Implementation of Machine Learning on ESP32-Based
Edge Devices 144
   Dalibor Dobrilovic

Stock Open Price Prediction of Software Companies in the BSE SENSEX
50 Index 156
   Chhaya Sonar and Ahmed M. Al Hammadi

Implementation of Morphological Fractional Order Darwinian Operator
for Brain Tumour Localization 169
   Kwabena Ansah, Wisdom Benedictus Adevu,
   Joseph Agyapong Mensah, and Justice Kwame Appati

Author Index 183

Intelligent Systems

Investigating Freshmen Students' Coding Standards Challenges Using NLP Techniques

Edgar Ceh-Varela[✉] and Essa Imhmed

Department of Mathematical Sciences, Eastern New Mexico University, Portales, NM 88130, USA
{eduardo.ceh,essa.imhmed}@enmu.edu

Abstract. This study investigates the potential use of Natural Language Processing (NLP) techniques to analyze coding standards violations within the context of an introductory programming course. In particular, the study evaluates the effectiveness of various advanced text embedding techniques, including Bag of Words (BOW), Doc2Vec, and BERT, in clustering coding standards violations. This study aims to determine which text embedding techniques contribute to the most accurate clustering of errors. Our findings highlight the superiority of Doc2Vec embeddings in effectively clustering related errors compared to the alternative techniques.

Keywords: Coding Conventions · Static Code Analysis · Text Embeddings · Clustering

1 Introduction

Coding Standards are guidelines and best practices emphasized in the software industry to produce high-quality code [1]. Adherence to such standards results in writing code improved for readability and, hence, enhancing communication and collaboration among software teams, as every developer comprehends code written by their peers. Consequently, this reduces software maintenance costs and the risk of software bugs, increasing the likelihood of meeting software delivery dates.

Code quality is also essential in an academic environment. It promotes student collaboration in class projects, as every team member can understand the code written by their peers, enhancing the overall project experience. Furthermore, when students submit code with good quality, instructors could promptly review their code for grading or debugging rather than spending effort and time attempting to understand poorly written code. Nevertheless, students often neglect to adhere to coding standards, justifying it by having tight school schedules, which implies a prioritization of code functionality over quality [1–5].

Several studies employed various techniques to analyze code quality issues among college students. In [6], the authors utilized static analysis tools, such as CheckStyle [7] and PMD [8], to uncover code quality violations among novice undergraduates. Moreover, in [9], the authors explore the feasibility of using Natural Language Processing (NLP) techniques to automate code duplicate detection. Recently, researchers have used

NLP to cluster code style errors and gain insights into areas needing focus through code demonstrations and conceptual coding standards assignments [10].

In this study, we identify code quality issues—focusing mainly on coding conventions—among first-year computer science undergraduates using CheckStyle and PMD tools for Java programming language assignments. We also compare text embedding methods—BOW [11], Doc2Vec [12], and BERT [13] in particular— to analyze if the generated embeddings can be used to cluster similar coding standards violations. The resulting clusters provide insight into students' fundamental challenges when learning Java syntax, logic, and problem-solving. These standard error groupings can guide curriculum design and adaptive feedback. For further details on how and why the NLP techniques are used, please refer to Sect. 3.

The remainder of this paper is organized as follows. Section 2 presents some background related to our research. Section 3 describes the proposed method in detail. We present the results of this study in Sect. 4. Finally, conclusions and future work are presented in Sect. 5.

2 Background

Recent studies investigate code quality in an academic environment. Li et al. [1] explored various teaching strategies to enhance students' perception of code quality. These strategies include assigning a small set of common coding standards suitable for novice programmers, providing students with a document outlining coding standards expectations, and incorporating practical exercises related to coding standards. Li et al. also recommend implementing peer assessment to convey the importance of code quality among students. Stegeman et al. [14] incorporated grading rubrics enforcing code quality in student code submissions into their programming courses.

Other studies [6, 15] have examined static analysis tools, such as CheckStyle and PMD, to identify code quality issues among novice CS majors and non-CS undergraduates in introductory programming courses. Their results revealed common errors related to formatting and documentation regardless of their programming skills and degree major. Likewise, in [16], the researchers compared three open-source static analysis tools for Java programs: PMD, FindBugs, and CheckStyle. Their findings show that each tool uncovers distinct types of bugs. PMD identifies common programming flaws, such as empty catch blocks and unused variables. FindBugs detects bugs and potential runtime issues, while CheckStyle focuses on code style violations.

Text similarity measurement is crucial in NLP tasks [17]. Measuring semantic similarity, especially in short texts, presents challenges due to their unique characteristics, such as limited context, concatenation of noun phrases, oral expressions, spelling errors, and ambiguity [18, 19]. Methods for short text similarity can be categorized as word- and semantic-based. BOW is a word-based method that represents text by counting the occurrence of individual words in a document, focusing on word frequency. Semantic-based methods encode short texts into continuous vectors using word embeddings, representations of words in an n-dimensional space [20, 21]. Doc2Vec and BERT are semantic-based methods that capture document-level semantics; Doc2Vec generates continuous vectors for entire documents. BERT employs transformer-based neural networks to produce contextualized embeddings, considering the meaning of words in the context of the entire document.

3 Methodology

3.1 Data Collection

Table 1. Coding standards criteria from the grading rubric

Criteria	Ratings
Coding Standards	**10 pts Excellent (100%)** Clearly and effectively documented, including the JavaDoc tags for the author's name, date, and assignment title; includes descriptive names of all program variables and functions; specific purpose noted for each function, control structure, input requirements, and output results; excellent use of white space; creatively organized work
	6 pts Satisfactory (60%), but needs Improvement It includes JavaDoc tags for the author's name, date, and assignment title; basic documentation has been completed, including descriptive names of all program variables and functions; purpose is noted for each function; white space makes the program fairly easy to read; organized work
	4 pts Unsatisfactory ($\leq$40%) No author name, date, or assignment title is included; very limited or no documentation is included; documentation and naming do not help the reader understand the code; poor use of white space (indentation, blank lines); disorganized and messy

We integrated code quality into an introductory Java programming course at Eastern New Mexico University (ENMU). Like [1, 14], we introduced our students to a small set of coding standards suitable for novice programmers yet typical for completing medium-complexity programs. These standards focus on conventions related to documentation/comments, size violations (e.g., line length), naming conventions, white space, and code block violations. In addition, we developed a grading rubric to assess students' problem-solving abilities and algorithmic thinking of programming assignments and enforce code quality in their code submissions. Table 1 shows the coding standards criteria from the rubric.

We collected 507 Java programming assignment solutions submitted by students over a three-semester period from Fall 2022 through Fall 2023. We used CheckStyle and PMD to perform a static analysis of code submissions. Like [6], we used CheckStyle to check source code for coding conventions and style and PMD to identify non-simplified expressions and code issues that may lead to potential software bugs. We configured CheckStyle and PMD to check for violations of the coding standards set in the grading rubric. Subsequently, we obtained reports detailing more than 27,000 instances of coding standards violations. Each instance includes a file name and a code line number indicating where the violation has occurred. It also describes the violation and its category.

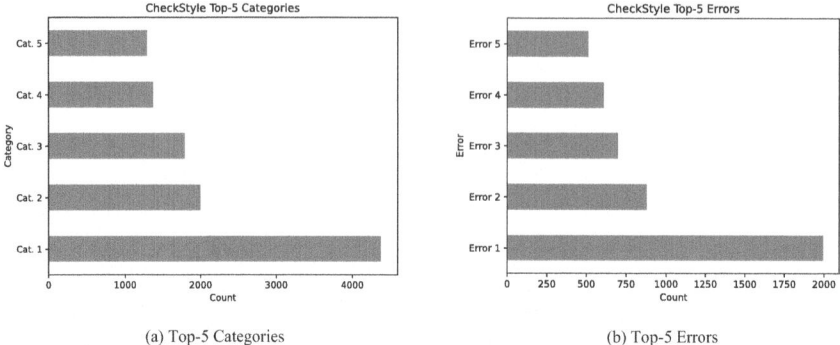

(a) Top-5 Categories (b) Top-5 Errors

Fig. 1. Top Categories and Errors from CheckStyle. The top categories in order are: "WhitespaceAround," "RegexpSingleline," "LineLength," "LeftCurly," "FinalParameters." The top errors in order are: "Line has trailing spaces," " 'if' is not followed by whitespace," " '{' is not preceded with whitespace," "Missing a Javadoc comment," "Missing package-info.java file."

For CheckStyle, we obtained 16,851 instances of errors. These errors belong to 49 categories. Figure 1 shows the top-5 most common categories and errors from the CheckStyle dataset. The most frequent category is WhiteSpaceAround[1]. This category refers to improper use of whitespaces around operators, keywords, and blank lines, affecting the code readability. The most common error in our dataset is "Line has trailing

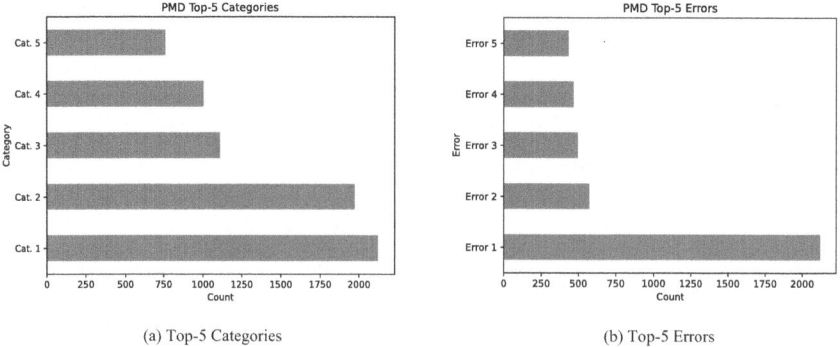

(a) Top-5 Categories (b) Top-5 Errors

Fig. 2. Top Categories and Errors from PMD. The top categories in order are: "SystemPrintln,""LocalVariableCouldBeFinal," "CommentRequired," "ShortVariable," and "MethodArgumentCouldBeFinal." The top errors in order are: "Usage of System.our/err," "Public method and constructor comments are required," "This utility class has a non-private constructor," "Class comments are required," "Ensure that resources like this Scanner object are closed after use."

---

[1] https://checkstyle.org/checks/whitespace/whitespacearound.html.

spaces." This error in Java coding standards refers to having white space characters like spaces or tabs at the end of a line[2]. Trailing white space in code is considered bad practice because it makes it look messy and inconsistently formatted.

For PMD, we obtained 10,347 instances of errors. These errors belong to 60 categories. Figure 2 shows the top-5 most common categories and errors from the PMD dataset. The most frequent category is SystemPrintln[3]. This category refers to the use of print statements for logging. The most common error in our dataset is "Usage of System.out/err." This error in Java coding standards refers to using 'System.out' or 'System.err' for logging purposes may not be the best practice.

3.2 Data Preprocessing

The reports from CheckStyle and PMD contain multiple text entries with the standard errors found on each student's assignment. The errors contain information such as variable names, digits, or operator symbols that could be more informative for our study. We preprocess the error descriptions to create a cleaner dataset.

We created several representative placeholders to replace specific tokens from our dataset. For example, we use the $<$ A CLASS NAME $>$, $<$ A METHOD NAME $>$, and $<$ BRACKET $>$ to represent the presence of a specific class name, a specific method, and any brackets (i.e., {, [, (,),], }) present in the error descriptions. Using abstract placeholders instead of specific names makes the text applicable to various scenarios. This transformation allows the concepts to be discussed more generally. Similarly, the structured placeholders allow easy finding and replacing operations instead of trying to locate random class names. Finally, we transformed all the text to lowercase characters. The preprocessing allows us to transform errors like "Parameter args should be final" and "'*' is not followed by whitespace" to "parameter $<$ a parameter name $>$ should be final" and " $<$ operator $>$ is not followed by whitespace," respectively.

Figure 3 shows the top 10 tokens that appear in our datasets after the initial preprocessing. Syntax errors, whitespace issues, and operator usage are among the most common coding errors in the dataset. The word cloud gives insight into the predominant categories of errors made by the students at this level based on the frequency.

Fig. 3. Word Cloud Showing the Top 10 Words After the Preprocessing

---

[2] https://www.oracle.com/java/technologies/javase/codeconventions-whitespace.html.
[3] https://pmd.github.io/pmd/pmd_rules_java_bestpractices.html#systemprintln.

3.3 Text Embeddings

Text embeddings [22] are numerical representations of words or phrases in a high-dimensional space, where the distance and direction between vectors capture semantic relationships between the words. This study compares three techniques to produce text embeddings: Bag Of Words (BOW), Doc2Vec, and BERT.

Bag of Words. The Bag of Words (BOW) embedding technique is a representation method used in NLP to convert text data into numerical vectors. In this approach, a document is represented as an unordered set of words, disregarding grammar and word order but considering their frequency of occurrence. The process involves creating a vocabulary of unique words in the entire corpus and then constructing vectors for each document based on the frequency of these words. Each dimension in the vector corresponds to a unique word, and the value in that dimension represents the frequency of the word in the document. As a result, we obtained a sparse representation of the text, where the order of words is neglected, making it a simple yet effective method for text analysis tasks like document classification and sentiment analysis.

After applying this technique to our datasets, we obtained an embedding size of 123 for CheckStyle and 309 for PMD. As mentioned, these dimensions correspond with the vocabulary size for each dataset.

Doc2Vec. We use Doc2Vec, an unsupervised learning algorithm, to obtain embeddings for the error descriptions for both CheckStyle and PMD. In Doc2Vec, each document is represented as a fixed-size vector, regardless of length. Coding errors can be treated as short text "documents," making Doc2Vec appropriate for learning representations of these errors.

However, the dimensionality of the embeddings can impact the representation power of a Doc2Vec model. Larger embeddings have more parameters and dimensions to encode semantic information about documents. Smaller embeddings may be too limited. Therefore, we must find an optimal size that yields good performance without unnecessary complexity. To find this optimal size for the Doc2Vec dimensions for our datasets, we analyze the variance explained [23] when a given dimension size is reduced to only two dimensions. We use Principal Component Analysis (PCA) [23] as our dimensionality reduction technique. PCA is a statistical technique for dimensionality reduction and data visualization. By selecting a subset of these principal components, PCA allows for a lower-dimensional data representation while preserving the most important patterns and relationships. Therefore, the variance explained is the proportion of the total variance in the data captured or represented by a particular principal component or a combination of several principal components.

As dimensions increase, more semantic information can be encoded, increasing the variance explained. However, higher dimensions cause greater computational cost and risk of overfitting. We analyze 16, 32, 64, 128, 256, and 512 dimensions. Figure 4 shows elbow plots where we can see that after 256 dimensions, there is no significant change in the variance explained by our data. Therefore, we select 256 dimensions for the Doc2Vec embeddings on each dataset.

BERT. The Bidirectional Encoder Representations from Transformers (BERT) is a state-of-the-art pre-trained language model in NLP. BERT captures contextual information

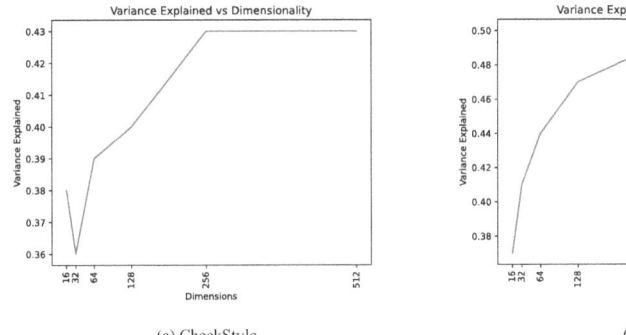

(a) CheckStyle (b) PMD

Fig. 4. Elbow Plot to Analyze the Variance Obtained From Different Embedding Dimensions

and word relationships by considering the entire surrounding context of each word in a bidirectional manner. Similarly, the model learns rich contextualized representations by training on large amounts of diverse text data. It employs a transformer architecture that captures intricate dependencies and nuances in language. BERT embeddings provide a more sophisticated and nuanced understanding of language, enabling the model to grasp meaning in context.

A pre-trained BERT model can be fine-tuned on specific tasks. Pre-trained models have a specific embedding size, indicating the dimensionality of the vector representations they generate for words or tokens. We use the pre-trained HuggingFace BERT-base-uncased model[4], which has a 768-dimensional embedding size.

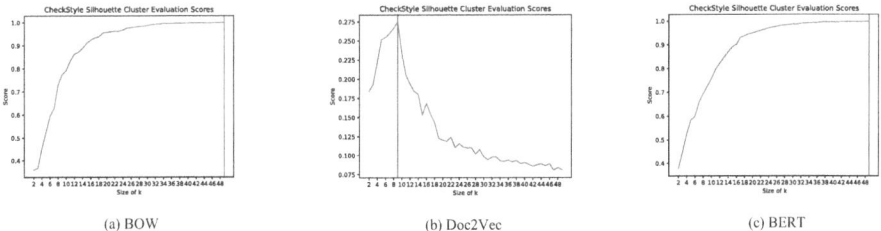

(a) BOW (b) Doc2Vec (c) BERT

Fig. 5. Silhouette Elbow Plot to Find the Optimal Value of k for CheckStyle

3.4 Clustering Errors

By clustering embeddings of programming errors, we can uncover common mistakes and misconceptions students encounter when learning to code. We use the k-means algorithm [24] on the document embeddings generated by BOW, Doc2Vec, and BERT to get the clusters. The k-means algorithm aims to partition observations into k clusters, such that each observation belongs to the cluster with the nearest mean, minimizing the

---

[4] https://hugginface.com/bert-base-uncased.

within-cluster variation. The number of clusters used in a k-means algorithm must be manually set. The Silhouette score [25] can help find the optimal number of clusters (i.e., k) for k-means clustering. The Silhouette score is a metric used to measure the goodness of a clustering technique. It provides a way to assess how well-separated the clusters are in the data. To compute the best value of k, we vary k to go from 2 to the number of categories in each dataset (e.g., for CheckStyle, it is 49, and for PMD, it is 60).

3.5 CheckStyle

Figure 5 shows the value of k for the CheckStyle dataset using the highest Silhouette score for each embedding technique. For BOW, the best value for k is 49, indicating that there must be a cluster for each of the 49 categories in the CheckStyle dataset. For Doc2Vec, the best value for k is 9. This value shows that Doc2Vec can group different errors in a cluster. Finally, for BERT, the best value for k is also 49, meaning each error should be in an individual cluster.

3.6 PMD

Figure 6 shows the value of k for the PMD dataset using the highest Silhouette score for each embedding technique. For BOW, the best value for k is 60, indicating that there must be a cluster for each of the 60 categories in the PMD dataset. For Doc2Vec, the best value for k is 11. This value shows that Doc2Vec can group different errors in a cluster. Finally, for BERT, the best value for k is 58, almost similar to the number of categories in the dataset.

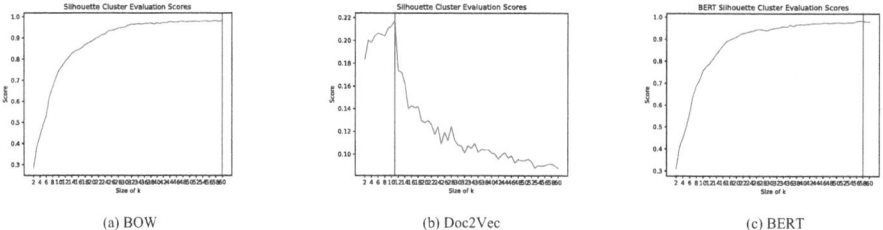

(a) BOW (b) Doc2Vec (c) BERT

Fig. 6. Silhouette Elbow Plot to Find the Optimal Value of k for PMD

4 Results

In this section, we present the results obtained by clustering the error for each dataset using the three embedding methods.

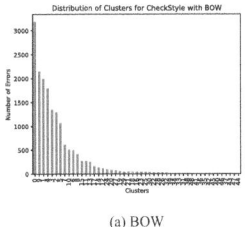 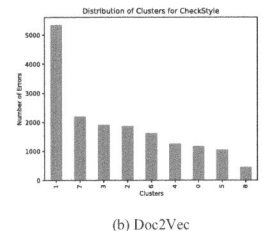 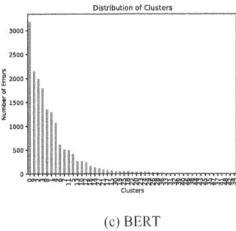

(a) BOW　　　　　　　　　(b) Doc2Vec　　　　　　　　　(c) BERT

Fig. 7. Errors distribution for CheckStyle

4.1 CheckStyle

Using the value of k obtained in the previous section, we clustered the errors and verified if the generated clusters were well formed and if they captured the meaning of the errors correctly.

Figure 7 shows the error distributions for the CheckStyle dataset for each embedding method using the value of k obtained previously. The Figure shows that using BOW and BERT produces a long-tailed distribution, indicating that only a few errors fall in those clusters. However, we can see that Doc2Vec groups errors in fewer clusters, considering the semantic meaning of these errors.

We evaluate the Silhouette score for each cluster to validate how good the clusters are. Figure 8 shows the Silhouette score for each cluster using the different embedding techniques. We can see that BOW and BERT produce dense clusters, as there is one cluster for each of the 49 categories in the CheckStyle dataset. Moreover, we can see that Doc2Vec also produces good clusters even if there are only 9 clusters. These results confirm that CheckStyle using Doc2Vec can group related errors.

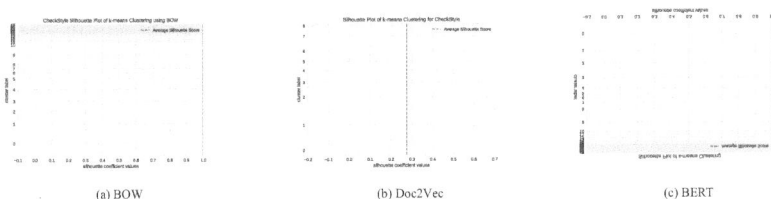

(a) BOW　　　　　　　　　(b) Doc2Vec　　　　　　　　　(c) BERT

Fig. 8. Silhouette Score for each cluster for CheckStyle

Figure 9 shows part of the content of a cluster using Doc2Vec on CheckStyle. We can see that Doc2Vec can group different errors with similar meanings, such as missing whitespace before or after an operator.

```
<operator> is not preceded with whitespace
 <operator> is not followed by whitespace
<operator> is not preceded with whitespace
 <operator> is not followed by whitespace
<operator> is not preceded with whitespace
```

Fig. 9. Similar errors clustered using Doc2Vec for CheckStyle

4.2 PMD

We analyzed the results obtained for the PMD dataset. We clustered the errors and verified if the generated clusters were well formed and correctly captured the errors' meaning.

Figure 10 shows the error distributions for the PMD dataset for each embedding method using the value of k obtained previously.

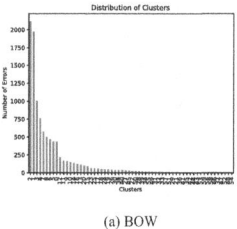
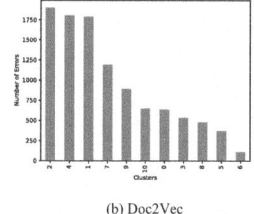
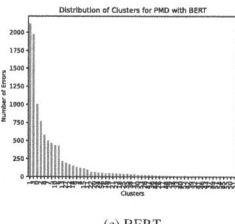

(a) BOW (b) Doc2Vec (c) BERT

Fig. 10. Errors distribution for PMD

As we can see, similar to the results obtained for CheckStyle, using BOW and BERT produces a distribution with long tails. Doc2Vec groups errors in fewer clusters, considering the semantic meaning of these errors. Moreover, Fig. 11 shows part of the content of a cluster using Doc2Vec on PMD. The Figure shows that Doc2Vec can group errors with similar meanings. For example, avoiding short names for classes or variables.

```
avoid variables with short names like <a_varia...
     avoid short class names like <a_class>
avoid variables with short names like <a_varia...
avoid variables with short names like <a_varia...
avoid variables with short names like <a_varia...
```

Fig. 11. Similar errors clustered using Doc2Vec for PMD

We evaluate the Silhouette score for each cluster generated for the PMD dataset. Figure 12 shows the Silhouette score for each cluster using the different embedding techniques. We can see that BOW and BERT produce dense clusters, as there are 60 and 58 clusters, respectively. We need to recall that there are 60 categories in the PMD dataset. Therefore, no major grouping of errors is created. Moreover, we can see that Doc2Vec also produces good clusters even if there are only 11 clusters.

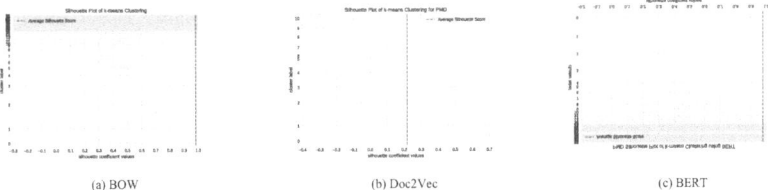

Fig. 12. Silhouette Score for each cluster for PMD

The results confirmed that Doc2Vec can group errors with similar meanings, such as avoiding short names for classes or variables.

5 Conclusions

The implementation of coding standards is crucial within the software industry. We incorporated these standards into the curriculum of an introductory programming course, utilizing CheckStyle and PMD outputs to identify violations of coding conventions in Java programming assignments. Employing BOW, Doc2Vec, and BERT embeddings, we clustered errors to uncover broader concepts and challenges encountered by students in their coding journey. Our findings reveal that, although not perfect, Doc2Vec embeddings excel in clustering related errors compared to the other two techniques, providing a more precise representation of novice programmers' challenges. For instance, students often omit optional white space, adversely impacting code readability, as supported by the findings in [1, 15].

Our future work includes exploring alternative teaching strategies to enhance students' awareness of and adherence to coding standards. Additionally, we aim to automate coding standards assessment, allowing students to receive prompt, constructive feedback on code quality well before final submissions, reducing instructors' efforts, and providing insights into students' code quality. We intend to implement an LLVM compiler extension [26–29] for detecting code clones and best practices violations. Furthermore, we plan to employ machine learning techniques [30] to predict students' performance and learning outcomes.

References

1. Li, X., Prasad, C.: Effectively teaching coding standards in programming. In: Proceedings of the 6th Conference on Information Technology Education. SIGITE 2005, pp. 239–244. Association for Computing Machinery, New York, NY, USA (2005). https://doi.org/10.1145/1095714.1095770
2. Chen, H.-M., Chen, W.-H., Lee, C.-C.: An automated assessment system for analysis of coding convention violations in java programming assignments. J. Inf. Sci. Eng. **34**, 1203–1221 (2018)
3. Hofbauer, M., Bachhuber, C., Kuhn, C., Steinbach, E.: Teaching software engineering as programming over time. In: 2022 IEEE/ACM 4th International Workshop on Software Engineering Education for the Next Generation (SEENG), pp. 51–58 (2022). https://doi.org/10.1145/3528231.3528353

4. Karnalim, O., Simon, Chivers, W.: Work-in-progress: code quality issues of computing undergraduates. In: 2022 IEEE Global Engineering Education Conference (EDUCON), pp. 1734–1736 (2022). https://doi.org/10.1109/EDUCON52537.2022.9766807
5. Karnalim, O., Simon: Promoting code quality via automated feedback on student submissions. In: 2021 IEEE Frontiers in Education Conference (FIE), pp. 1–5 (2021). https://doi.org/10.1109/FIE49875.2021.9637193
6. Albluwi, I., Salter, J.: Using static analysis tools for analyzing student behavior in an introductory programming course. Jordanian J. Comput. Inform. Technol. (JJCIT) **6**(3), 215–233 (2020)
7. Checkstyle. https://checkstyle.sourceforge.io. Accessed 7 Jan 2024
8. PMD. https://pmd.github.io. Accessed 7 Jan 2024
9. He, J., Xu, L., Yan, M., Xia, X., Lei, Y.: Duplicate bug report detection using dual-channel convolutional neural networks. In: Proceedings of the 28th International Conference on Program Comprehension, pp. 117–127 (2020)
10. Imhmed, E., Ceh-Varela, E., Scott, K.: Identifying code quality issues for undergraduate students using static analysis and NLP. In: 2023 International Conference on Computational Science and Computational Intelligence (CSCI). IEEE (2023)
11. Abubakar, H.D., Umar, M., Bakale, M.A.: Sentiment classification: Review of text vectorization methods: bag of words, Tf-Idf, Word2vec and Doc2vec. SLU J. Sci. Technol. **4**(1 & 2), 27–33 (2022)
12. Le, Q., Mikolov, T.: Distributed representations of sentences and documents. In: International Conference on Machine Learning, pp. 1188–1196. PMLR (2014)
13. Devlin, J., Chang, M.-W., Lee, K., Toutanova, K.: BERT: pre-training of deep bidirectional transformers for language understanding. arXiv preprint arXiv:1810.04805 (2018)
14. Stegeman, M., Barendsen, E., Smetsers, S.: Designing a rubric for feedback on code quality in programming courses. In: Proceedings of the 16th Koli Calling International Conference on Computing Education Research. Koli Calling 2016, pp. 160–164. Association for Computing Machinery, New York, NY, USA (2016). https://doi.org/10.1145/2999541.2999555
15. Edwards, S.H., Kandru, N., Rajagopal, M.B.M.: Investigating static analysis errors in student java programs. In: Proceedings of the 2017 ACM Conference on International Computing Education Research. ICER 2017, pp. 65–73. Association for Computing Machinery, New York, NY, USA (2017). https://doi.org/10.1145/3105726.3106182
16. Oskouei, E.H., Kalıpsız, O.: Comparing bug finding tools for java open source software (2018)
17. Wang, J., Dong, Y.: Measurement of text similarity: a survey. Information **11**(9), 421 (2020)
18. Han, M., Zhang, X., Yuan, X., Jiang, J., Yun, W., Gao, C.: A survey on the techniques, applications, and performance of short text semantic similarity. Concurr. Comput. Pract. Exp. **33**(5), 5971 (2021)
19. Prakoso, D.W., Abdi, A., Amrit, C.: Short text similarity measurement methods: a review. Soft. Comput. **25**, 4699–4723 (2021)
20. Selva Birunda, S., Kanniga Devi, R.: A review on word embedding techniques for text classification. In: Raj, J.S., Iliyasu, A.M., Bestak, R., Baig, Z.A. (eds.) Innovative Data Communication Technologies and Application. LNDECT, vol. 59, pp. 267–281. Springer, Singapore (2021). https://doi.org/10.1007/978-981-15-9651-3_23
21. Ceh-Varela, E., Imhmed, E.: Uncovering water research with natural language processing. In: 2023 IEEE 47th Annual Computers, Software, and Applications Conference (COMPSAC), pp. 983–984 (2023). https://doi.org/10.1109/COMPSAC57700.2023.00138
22. Mitra, B., Craswell, N.: Neural text embeddings for information retrieval. In: Proceedings of the Tenth ACM International Conference on Web Search and Data Mining, pp. 813–814 (2017)
23. Greenacre, M., Groenen, P.J., Hastie, T., d'Enza, A.I., Markos, A., Tuzhilina, E.: Principal component analysis. Nat. Rev. Methods Primers **2**(1), 100 (2022)

24. Sinaga, K.P., Yang, M.-S.: Unsupervised k-means clustering algorithm. IEEE Access **8**, 80716–80727 (2020)
25. Shahapure, K.R., Nicholas, C.: Cluster quality analysis using silhouette score. In: 2020 IEEE 7th International Conference on Data Science and Advanced Analytics (DSAA), pp. 747–748. IEEE (2020)
26. Imhmed, E., Cook, J., Badawy, A.-H.: Evaluation of a novel scratchpad memory through compiler supported simulation. In: 2022 IEEE High Performance Extreme Computing Conference (HPEC), pp. 1–7. IEEE (2022)
27. Imhmed, E.A.: Understanding performance of a novel local memory store design through compiler-driven simulation. PhD thesis, New Mexico State University (2022)
28. Akhila, C., Saleena, N.: Value based redundancy detection in SSA code. In: 2016 IEEE Annual India Conference (INDICON), pp. 1–5. IEEE (2016)
29. Zhang, M.: Detecting redundant operations with LLVM. http://james0zan.github.io/resource/GSoC15-Proposal-BloatDetection.pdf. Accessed 10 April 2024
30. Abu-gellban, H., Zhuang, Y., Nguyen, L., Zhang, Z., Imhmed, E.: CSDLEEG: identifying confused students based on EEG using multi-view deep learning. In: 2022 IEEE 46th Annual Computers, Software, and Applications Conference (COMPSAC), pp. 1217–1222 (2022). https://doi.org/10.1109/COMPSAC54236.2022.00192

ParkinSense: A Tri-Sensory AI-Powered Framework for Early Detection of Parkinson's Disease

M. Akshaya[1] and Emmanuel Joy[2(✉)]

[1] Division of Computer Science and Engineering, Karunya Institute of Technology and Sciences, Coimbatore, India
makshaya@karunya.edu.in
[2] Division of Artificial Intelligence and Machine Learning, Karunya Institute of Technology and Sciences, Coimbatore, India
dremmjoy@outlook.com

Abstract. Parkinson's disease (PD) is a brain dysfunction condition that affects thousands of people across the globe. Early detection of Parkinson's disease is crucial for effective treatment and control of symptoms. This study employs a multimodal methodology coded as *ParkinSense* that combines spiral drawings, gait analysis, and Magnetic Resonance Imaging (MRI) analysis for timely Parkinson's disease identification. MRI analysis provides insight into structural and functional changes in the brain, revealing potential anomalies. GAIT analysis assesses alterations in walking patterns, such as reduced arm swing and slow gait velocity. Spiral drawings, serve as a surrogate for motor dexterity and fine motor control, both of which are compromised in Parkinson's disease. In this study, data from Parkinson's patients and healthy candidates are collected and analyzed using advanced machine-learning techniques to create models for accurate disease prediction and classification. Specifically, the gait analysis is conducted using a Support Vector Machine and XGBoost algorithm. The spiral drawings are analyzed using a Convolutional Neural Network. The MRI data is processed using a MobileNet-based neural network. These algorithms are integrated to create models for accurate disease prediction and classification. The results demonstrate that this multimodal technique is equipped to identify Parkinson's disease with an accuracy rate of more than 90%. Early detection using speech analysis, gait analysis, and spiral drawings can lead to earlier intervention and treatment, resulting in a better excellence of life for those affected. This study adds to the expanding body of knowledge about non-invasive, cost-effective approaches for the timely identification of Parkinson's disease.

Keywords: Parkinson · MRI analysis · gait analysis · spiral drawings · machine learning · classification · clinical assessment

1 Introduction

Parkinson's disease is a dysfunction that affects innumerable individuals worldwide [1, 7]. It is crucial to determine the condition to initiate treatments and improve patients' quality of life. One way to identify Parkinson's disease in its initial phase is by analyzing voice, gait and spiral drawings. In effect, medical practitioners can implement effective measures to minimize the long-term impact of the disease.

MRI analysis [15] has emerged as a non-invasive and complete method to detect Parkinson's disease. Structural and functional changes within the brain, including alterations in neural connectivity, substantia nigra morphology, and overall brain atrophy, are elucidated through MRI analysis. Researchers developed algorithms to discern subtle differences, providing accurate differentiation between individuals with Parkinson's and those without. The use of MRI analysis holds potential for efficient screening in various healthcare settings, offering valuable insights into the neuroanatomical markers associated with Parkinson's disease.

Gait analysis [3] exhibits potential in detecting Parkinson's disease, a condition impacting the motor system. It induces walking irregularities such as shortened steps, shuffling gaits, and reduced walking speeds. Researchers used sensors or motion capture devices with markers to measure and analyze aspects of gait using machine learning. By doing they developed algorithms that differentiate between those diagnosed with Parkinson's disease and those without it. This provides an objective approach for the detection of the disease. An advantage of gait analysis is that it is non-invasive, affordable and easily applicable, in clinical settings.

Spiral drawings [4] have been identified as a conceivable diagnostic resource for Parkinson's disease identification, in addition to voice and gait analysis. Motor control issues are frequently found in those individuals dealing with Parkinson's disease which can contribute to poor handwriting. This can be seen in their spiral pattern, which is narrow in size, less fluid, and contains tremors. Utilizing image processing techniques and machine learning algorithms researchers can accurately identify between individuals diagnosed with Parkinson's disease and healthy individuals by extracting quantitative features such as curvature, speed and variability. Spiral drawings offer a quick way to evaluate motor control and they can be a screening tool in situations where people get basic care.

Finally, the integration of MRI analysis, gait analysis and spiral drawing evaluation has the potential to transform the early detection of PD. These non-invasive, cost-effective technologies provide objective indicators that can assist in identifying individuals who are in the initial progression of the disease or who are at risk of developing it. These approaches identify those afflicted by Parkinson's with the capacity for improved well-being by allowing for early interventions like medication and physical therapy. To guarantee their widespread application and integration into standard clinical practice more research and validation of these strategies are required.

The highlights of the proposed work *ParkinSense* are as follows:

1. Built sensory analysis models using machine learning for MRI analysis, gait analysis and spiral drawing to identify distinctive patterns for Parkinson's.
2. Used an ensemble framework for integration of tri-sensory models to increase the overall diagnostic accuracy.
3. Robust system tested across a wide range of patients and datasets, leading to a potential clinical diagnostic tool for healthcare systems.

The subsequent sections of this paper are organized as follows: The review of literature is outlined in Sect. 2. The methodology of the proposed system is presented in Sect. 3. The discussions and potential applications are emphasized in Sect. 4. The conclusion and scope of this work are summarized in Sect. 5.

2 Related Work

A large number of artificial intelligence-based methods are used in the identification of Parkinson's disease. This section presents a few of the recent methodologies involved in the process. Pahuja et al. [1] did a comparative study of existing machine-learning approaches for PD diagnosis and found that Levenberg–Marquardt algorithm was found to be the best classifier, having the highest classification accuracy of 95.89%. Liaqat Ali et al. [2] proposed the phonation-based LDA-NN-GA model, which has issues with gender distribution but achieves 100% testing and 95% training accuracy. This method excels in PD detection focused solely on acoustic voice analysis. Luca Parisi suggested the m-ark-SVM [3] method which enhances Support Vector Machines (SVMs) with the m-arcsinh kernel for speech feature recognition. More training data is needed to reduce the overlapping areas so that it can accurately classify speech patterns associated with Parkinson's disease. Kamble et al. [4] proposed a digitized spiral drawing classification for PD diagnosis. It achieves 91.6% accuracy in distinguishing PD patients but has fewer samples tested to prove its real-time efficiency. Shetty et al. [5] proposed the SVM-based gait analysis model effectively distinguishes PD with 83.33% accuracy, but encounters specificity challenges.

Kemal [6] proposed a hybrid system that uses the FFT algorithm and logistic linear classifier to identify Parkinson's disease with a high accuracy of 81.3%. However, this method can be used for PD detection based on only one sensor. Wang et al. [7] presented a deep-learning approach that offered superior performance compared to machine learning models but limited datasets (183 healthy, 401 PD patients) raise generalizability concerns. Shu et al. [8] utilized a thirteen-layer CNN for precise early PD diagnosis but the emphasis on EEG signals limited their applicability. Quan et al. [9] explored Bidirectional LSTM models for PD detection. However, the proposed model focused on stage classification, limiting its exploration to multilabel classification tasks. Memedi et al. [10] presented an automatic spiral analysis method to objectively assess motor symptoms in Parkinson's disease. They created algorithms that examine spiral drawing patterns and identify traits associated with motor deficits. The study revealed how beneficial this approach is for quantitative disease evaluation.

Kisor Wagh et al. [11] used spiral and wave sketching patterns to identify Parkinson's by creating classification algorithms that examine these patterns, demonstrating the potential for accurate diagnosis. Kamble et al. [4] proposed digitized spiral drawings to diagnose Parkinson's disease. They created a classification algorithm that uses digital information from the drawings to discriminate between healthy individuals and Parkinson's patients. Mary et al. [12] explored ways to recognize Parkinson's disease through voice cues and sketching patterns. They assessed the effectiveness of various methodologies and emphasized the significance of multimodal analysis for correct diagnosis. Pham et al. [13] suggested a multimodal strategy for detecting Parkinson's disease using speech analysis and an improved spiral test. They used machine learning techniques to show how good their method is for accurately detecting Parkinson's disease. In their study, Rao et al. [14] used a dataset of voice recordings and spiral drawings to detect Parkinson's illness using a classification algorithm that uses various multimodal data sources to make reliable diagnoses.

The current study on Parkinson's disease neglects the precision gained from multimodal integration, favouring single modalities. For improved diagnostic precision, ParkinSense offers a tri-sensory technique that combines MRI, gait, and spiral drawing evaluations. For broad early detection across demographics, ParkinSense uses ensemble approaches to solve dataset restrictions and generalizability. Even with excellent diagnostic accuracy, practical application issues such as clinical integration and cost remain. ParkinSense has an emphasis on price, accuracy, clinical integration, and accessibility to facilitate the implementation of sophisticated diagnostics in routine clinical settings.

3 Proposed Methodology

3.1 Multimodal Data Processing Module

The multimodal data processing module's primary objective is based on the thorough processing and analysis of diverse data inputs, which serve as a pivotal component of *ParkinSense*. The inputs span CSV files containing gait data, spiral drawings and MRI captured as image files. Several key aspects are encompassed within this module's functionality. Firstly, the module applies sophisticated data preprocessing methods tailored to each modality. This includes cleaning and formatting data to optimize quality and consistency. Robust techniques are utilized to address potential noise, outliers, or irregularities inherent in diverse data sources. Moreover, the module incorporates advanced algorithms for feature extraction mechanisms specific to GAIT data, and image files. These techniques are intended to distil relevant information and patterns from the unprocessed data, making the subsequent stages of analysis easier. The extracted features are subjected to normalization and standardization processes to ensure uniformity and comparability across different modalities. This crucial phase enhances the multimodal dataset's overall reliability and cohesion, contributing to the efficiency of subsequent analytical procedures. The module incorporates error-checking mechanisms capable of handling missing or incomplete data to strengthen the system against potential challenges. This promotes a robust and resilient data analysis framework by safeguarding

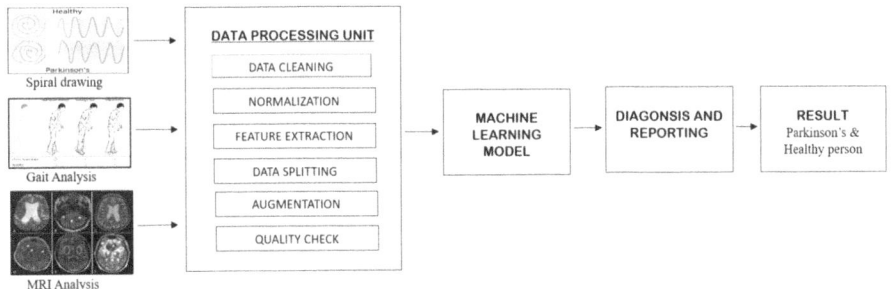

Fig. 1. The design of the envisioned system – ParkinSense.

against disruptions in the processing pipeline. By meticulously preparing and refining the data, the multimodal data processing module ultimately serves as a gateway to insightful analysis. When determining the probability of Parkinson's disease based on multimodal inputs, its output which consists of the standardized and processed features interoperates with the prediction module with ease (Fig. 1).

3.2 Machine Learning Prediction Module

Machine Learning prediction module stands as a pivotal component within the Parkinson's disease detection system and uses sophisticated predictive analytics models to predict the likelihood of developing Parkinson's disease. This predictive analysis is based on well-processed data derived from diverse multimodal inputs, such as MRI data, GAIT data and spiral drawings. At its core, the module leverages pre-trained machine learning models that make use of MRI analysis, GAIT analysis, and image classification for spiral drawings. These models can discern patterns and associations within their respective modalities because they were previously trained on pertinent information. The module implements an ensemble method to enhance the overall accuracy and reliability of predictions. With this strategic approach, predictions from different models are combined to create a synergistic effect that captures subtle information from each modality. The accuracy of Parkinson's disease detection is ultimately improved by the ensemble method which contributes to a more robust and comprehensive analysis. In addition to providing accurate predictions, the module prioritizes transparency and user understanding. To achieve this, it integrates explain ability techniques, offering users insights into the factors that significantly influence the model predictions. This interpretability aspect enhances the system's trustworthiness and empowers users to make informed decisions based on the generated predictions.

3.3 User Interaction Module

The User Interaction Module within the *ParkinSense* is built on the Streamlit web application (https://streamlit.io/) which plays a crucial role in elevating the user experience and facilitating seamless interaction. The primary objective of this module is to furnish an easily navigable and information-packed platform for users. To achieve this, improvements in the layout and design of the Streamlit app are initiated which focuses on creating a good user interface. Enhancing overall usability and simplifying navigation ensures that users can easily navigate through the app's functionalities. In an attempt to provide real-time data input options, the module incorporates features that let users upload MRI images, gait data, or spiral drawings directly through the app. This enables a more interactive and dynamic user experience while simplifying the input data processing (Figs. 2 and 3).

The key focus of the enhanced user interface is to show the all-inclusive results. Access to the complete data, such as model outputs, confidence scores, and explanations for the forecasts, is guaranteed by the module. This transparency empowers users with a deeper understanding of how the system makes decisions, which promotes trust and well-informed choices. The module enables users to choose between different models or modalities for analysis as per user preferences and demands. This allows users to

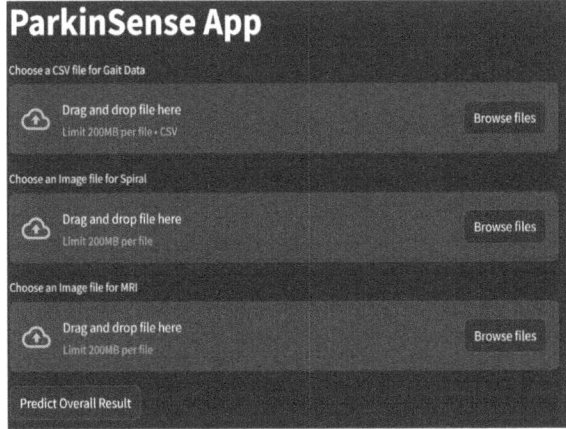

Fig. 2. The User Interface of the ParkinSense application

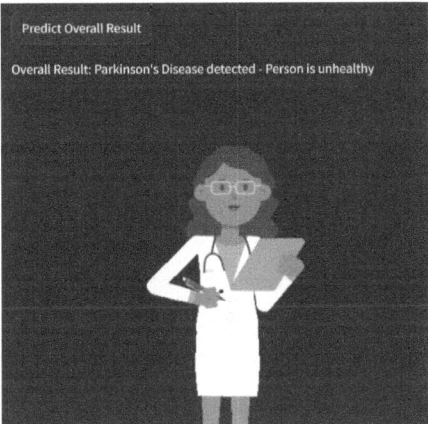

Fig. 3. Prediction results of the proposed system – ParkinSense

customize the analysis to meet their specific preferences, thus enhancing the overall versatility. The UI Module incorporates strong error-handling techniques to strengthen the user experience. To ensure a smooth and error-resistant interaction with the app these mechanisms provide informative messages to guide users through potential issues. This proactive approach addresses potential challenges in a user-friendly and supportive manner that contributes to user satisfaction.

In addition, various data formats and sources will be supported, which will enable more flexibility in data input. Performance indicators for improved model evaluation, like accuracy, will be also displayed. Deployment alternatives apart from Streamlit will be investigated in the future including cloud deployment for increased accessibility and scalability. Robust security measures will be implemented to ensure secure data handling, especially with sensitive medical data. In case of issues improved error-handling mechanisms will offer users meaningful messages, contributing to an overall positive user experience. The system's usability and transparency will be further enhanced with the introduction of comprehensive documentation, including user guides and system architecture details. Once implemented, these proposed enhancements will not only augment the application's capabilities but also address ethical and legal considerations in the context of health-related applications.

4 Results and Discussions

4.1 Datasets Used

To evaluate and test the proposed method, we employ the MRI dataset [16], the GaitPhase Database [17] for human gait analysis, and the spiral drawing dataset termed Parkinson's Drawings [18]. The brain MRI dataset uses structural features to diagnose Parkinson's disease. This uses data from 60 patients with Parkinson's disease and 56 healthy controls. An additional dataset of 69 patients with Parkinson's disease and 71 healthy controls was used for external validation. The gait dataset (Fig. 4) was created from 21 participants (10 males and 11 females) of normal age from 23.8 ± 3.3 years, a height of 172.8 ± 9.4 cm, and a weight of 66.6 ± 10.9 kg. The whole process involved a total of 25,306 steps comprising walking on a split-belt treadmill at twelve different speeds ranging between [0.6,1.7] m/s in increments of 0.1 m/s for one minute each at the given velocity. The spiral drawing dataset comprises images of spirals and waves obtained from healthy and PD patients. The data is further categorized into training and testing segments.

	L_FCC_x	L_FM1_x	L_FM2_x	L_FM5_x	R_FCC_x	R_FM1_x	R_FM2_x	R_FM5_x	L_FCC_y	L_FM1_y	...	R_FM2_y	R_FM5_y	L_FCC_z	L_FM1_z	L_FM2_z	L_FM5_z	R_FCC_z	R_FM1_z
0	1.127087	1.332534	1.328893	1.305142	0.804407	1.017605	1.017485	0.991847	0.575397	0.587244	...	0.448289	0.396237	0.090627	0.084655	0.095734	0.067779	0.081946	0.058241
1	1.131831	1.337421	1.333921	1.310700	0.802088	1.015296	1.015148	0.989509	0.575327	0.586249	...	0.448250	0.396193	0.090195	0.086384	0.097342	0.069010	0.082106	0.058203
2	1.136807	1.342521	1.339171	1.316510	0.799617	1.012833	1.012659	0.987012	0.575228	0.585174	...	0.448206	0.396145	0.089767	0.088228	0.099043	0.070312	0.082279	0.058166
3	1.141961	1.347770	1.344579	1.322506	0.797009	1.010229	1.010033	0.984370	0.575094	0.584023	...	0.448160	0.396093	0.089354	0.090171	0.100820	0.071672	0.082464	0.058132
4	1.147230	1.353103	1.350078	1.328615	0.794281	1.007500	1.007287	0.981602	0.574925	0.582801	...	0.448109	0.396038	0.088963	0.092198	0.102656	0.073079	0.082663	0.058101
...	...	...	...	...	...	...	...	...	...	...	...	...	...	...	...	...	...	...	...
11995	1.131665	1.334620	1.336033	1.312159	0.936247	1.150734	1.148909	1.122463	0.566691	0.562259	...	0.437999	0.387250	0.108844	0.074029	0.081974	0.055194	0.077946	0.061481
11996	1.136466	1.339685	1.341100	1.317486	0.934519	1.148988	1.147176	1.120718	0.566975	0.562450	...	0.437994	0.387244	0.107277	0.073884	0.082005	0.055195	0.077944	0.061450
11997	1.140658	1.344090	1.345509	1.322117	0.933004	1.147460	1.145657	1.119190	0.567223	0.562607	...	0.437990	0.387239	0.105965	0.073780	0.082054	0.055222	0.077943	0.061423
11998	1.144306	1.347908	1.349332	1.326128	0.931680	1.146127	1.144330	1.117854	0.567440	0.562735	...	0.437986	0.387234	0.104873	0.073708	0.082116	0.055269	0.077942	0.061400
11999	1.147532	1.351271	1.352702	1.329659	0.930503	1.144946	1.143152	1.116667	0.567631	0.562838	...	0.437983	0.387230	0.103952	0.073662	0.082189	0.055330	0.077940	0.061381

12000 rows × 24 columns

Fig. 4. GaitPhase Dataset

4.2 Baselines

ParkinSense delivers a robust and comprehensive solution for early diagnosis which is crafted through the implementation of multimodal data processing allowing users to analyze diverse inputs, such as MRI data, gait data, and spiral drawings. In the realm of GAIT data analysis, users can upload Comma comma-separated values (CSV) files to the system and receive predictions that show whether the patterns in the GAIT data are "Normal" or "Parkinsons". Similarly, users can also input spiral drawings images or

4.3 Performance Metrics

In *ParkinSense* accuracy metrics play the central role in evaluating the efficacy of each modal. In MRI analysis, accuracy serves as a crucial metric, assessing the model's proficiency in detecting distinctive neuroanatomical features indicative of Parkinson's disease, such as structural alterations, substantia nigra morphology, and overall brain abnormalities. Similarly for gait analysis accuracy measures the ability of the model to recognize patterns known as gait swings and arm swing variations which are the indicator that someone has PD. Additionally, in spiral drawings, accuracy determines whether the model can identify motor control problems specific to PD including small size dimensions, irregularity, and shakiness. Thus, accuracy emerges as the single metric for measuring performance across these modalities in determining whether they are successful at identifying early Parkinson's disease through their collective output.

4.4 Quantitative Analysis

For exploratory analysis, we plotted a histogram where subplots are arranged in a 5 × 5 grid (Fig. 5) to observe the distribution of a variable, from the gait dataset [16]. Each subplot consists of a histogram with the option to include a kernel density estimate. The title of each subplot indicates which feature variable is being visualized. The plot in

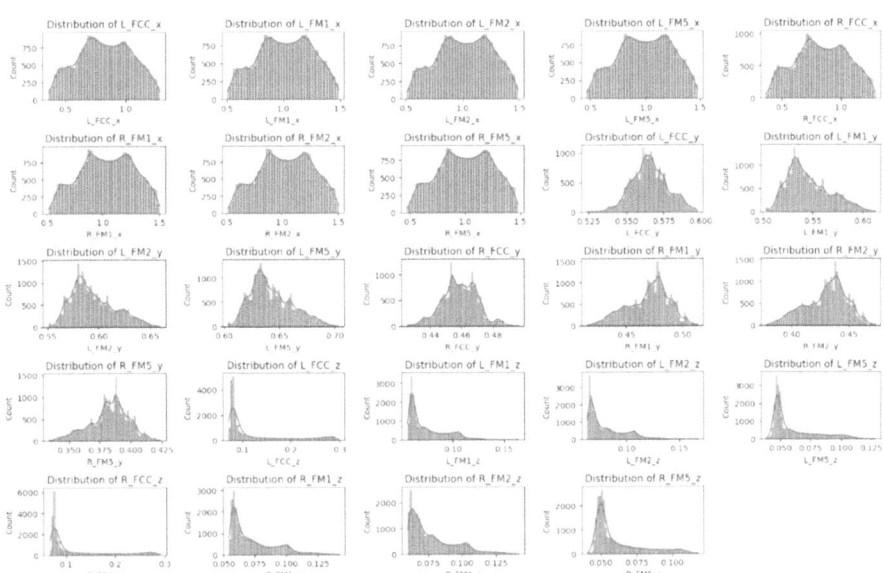

Fig. 5. Histogram with KDE during gait analysis

Fig. 6 shows the loss and accuracy of the MRI data analysis model from epoch to epoch. The model which is trained to provide an optimal solution continuously generalizes the training data. The plot in Fig. 7 demonstrates the changes in training accuracy and validation accuracy throughout the training process. By examining both curves, one can assess how effectively the model performs on data in comparison to its performance on the training set. If both curves show an increase and align with each other it suggests that the model is trained well and generalizes effectively. However, if the training accuracy continues to rise while the validation accuracy remains stagnant or decreases it could indicate overfitting (Table 1).

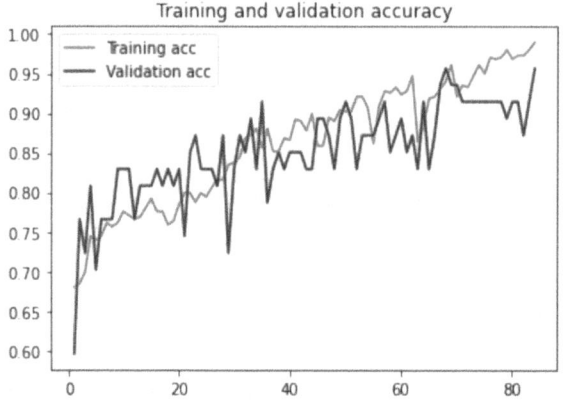

Fig. 6. Accuracy plot of MRI analysis

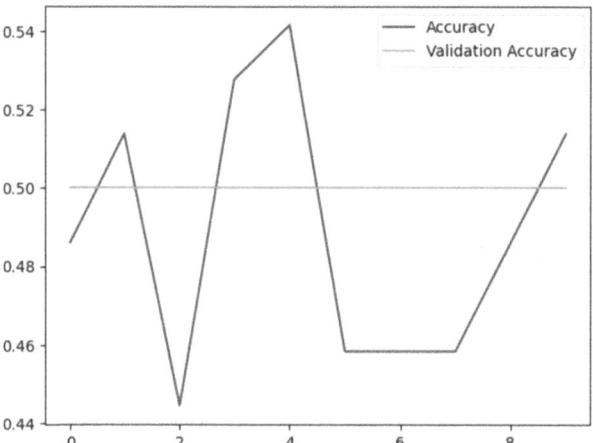

Fig. 7. Accuracy plot of spiral drawing during training

Table 1. Accuracy comparison results.

Algorithm	Existing model (Baseline)	Proposed model
Gait Analysis	0.85	0.93
Spiral Drawing Analysis	0.78	0.89
MRI Analysis	0.80	0.91
Ensemble (Combined model)	N/A	0.90

5 Conclusion and Future Work

The completion of the ParkinSense project represents a significant leap forward in the early diagnosis and management of Parkinson's disease. Using machine learning, multimodal data processing, and improved UI, the system provides a simplified tool to assess the chance of Parkinson's disease based on a range of inputs. The multimodal data processing module is crucial in preparing and enhancing data from spiral drawings, GAIT patterns, and MRI images. The system's ability to handle a wide range of data formats and extract relevant features guarantees a standardized and cohesive dataset for subsequent analysis. To enhance the overall accuracy of predictions, the module uses ensemble learning alongside pre-trained models for image, MRI, and GAIT analysis. Transparency and comprehension are promoted by the incorporation of appropriate techniques, which provide users with insights into the elements impacting model predictions. Through the Streamlit app, the UI is implemented, which not only makes the overall layout and design more user-friendly but also provides real-time data entry choices. Through the app users can conveniently upload voice, MRI, or GAIT patterns directly, contributing to a dynamic and engaging user experience. The new models achieve an overall accuracy of 90%, a significant improvement over current models, particularly in ensemble analysis. One of the main advantages of the project is its adaptability allowing users to customize analyses by choosing various models. Strong error-handling systems guarantee a flawless user experience. The *ParkinSense* is an important tool for physicians, researchers, and anyone else involved in Parkinson's disease diagnosis and treatment.

Future research into Parkinson's disease identification using MRI analysis, gait analysis, and spiral drawings can focus on a variety of areas to improve the system's usefulness and accuracy. To begin, research can be performed to investigate the integration of new indicators such as eye movements and tremor analysis, which may provide a more thorough multimodal assessment of Parkinson's disease. Furthermore, machine learning techniques can be developed to increase the system's predictive capabilities, allowing for earlier detection and action. The system can also be upgraded to include remote monitoring capabilities, allowing patients to receive regular assessments from the comfort of their own homes. This would not only make the study more convenient for patients, but it would also allow for a bigger pool of participants, thereby boosting the system's generalizability and scalability. Finally, longitudinal studies to assess the system's long-term performance and ability to follow illness development can provide useful information for clinical application and treatment monitoring. Overall, additional research and development in these areas can significantly increase Parkinson's disease detection and management. The open-source code of the project can be tested and evaluated from https://github.com/dremmjoy/ParkinSense.

References

1. Pahuja, G., Nagabhushan, T.N.: A comparative study of existing machine learning approaches for Parkinson's disease detection. IETE J. Res. **67**(1), 4–14 (2021)
2. Ali, L., Zhu, C., Zhang, Z., Liu, Y.: Automated detection of Parkinson's disease based on multiple types of sustained phonations using linear discriminant analysis and genetically optimized neural network. IEEE J. Transl. Eng. Health Med. **7**, 1–10 (2019)
3. Parisi, L., Ma, R., Zaernia, A., Youseffi, M.: M-ark-support vector machine for early detection of Parkinson's disease from speech signals. Int. J. Math. Comput. Simul. **15**, 34–41 (2021)
4. Kamble, M., Shrivastava, P., Jain, M.: Digitized spiral drawing classification for Parkinson's disease diagnosis. Meas. Sens. **16**, 100047 (2021)
5. Shetty, S., Rao, Y.S.: SVM-based machine learning approach to identify Parkinson's disease using gait analysis. In: 2016 International Conference on inventive computation technologies (ICICT), vol. 2, pp. 1–5. IEEE (2016)
6. Polat, K.: Freezing of gait (fog) detection using logistic regression in Parkinson's disease from acceleration signals. In: 2019 Scientific Meeting on Electrical-Electronics & Biomedical Engineering and Computer Science (EBBT), pp. 1–4. IEEE (2019)
7. Wang, W., Lee, J., Harrou, F., Sun, Y.: Early detection of Parkinson's disease using deep learning and machine learning. IEEE Access **8**, 147635–147646 (2020)
8. Oh, S.L., et al.: A deep learning approach for Parkinson's disease diagnosis from EEG signals. Neural Comput. Appl. **32**, 10927–10933 (2020)
9. Quan, C., Ren, K., Luo, Z.: A deep learning-based method for Parkinson's disease detection using dynamic features of speech. IEEE Access **9**, 10239–10252 (2021)
10. Memedi, M., et al.: Automatic spiral analysis for objective assessment of motor symptoms in Parkinson's disease. Sensors **15**(9), 23727–23744 (2015)
11. KishorWagh, M., RuchitaPawar, B., SoniyaChavan, V.: Parkinson's disease detection using spiral and wave drawing
12. Mary, G.P.A., Vignesh, G.N., Suganthi, N.: Various approaches to detecting Parkinson's disease using speech signals and drawing patterns. ICTACT J. Data Sci. Mach. Learn. **3**(01) (2021)
13. Pham, H.N., et al.: Multimodal detection of Parkinson's disease based on vocal and improved spiral test. In: 2019 International Conference on System Science and Engineering (ICSSE), pp. 279–284. IEEE (2019)
14. Rao, K.M.M., Reddy, M.S.N., Teja, V.R., Krishnan, P., Aravindhar, D.J., Sambath, M.: Parkinson's disease detection using voice and spiral drawing dataset. In: 2020 Third International Conference on Smart Systems and Inventive Technology (ICSSIT), pp. 787–791. IEEE (2020)
15. Heim, B., Krismer, F., De Marzi, R., Seppi, K.: Magnetic resonance imaging for the diagnosis of Parkinson's disease. J. Neural Transm. **124**(8), 915–964 (2017). https://doi.org/10.1007/s00702-017-1717-8
16. https://www.michaeljfox.org/news/parkinsons-progression-markers-initiative-ppmi
17. https://www.mad.tf.fau.de/research/activitynet/gaitphase-database/
18. https://www.frontiersin.org/journals/neurology/articles/10.3389/fneur.2017.00435

Analysing the Role of Post Click Factors in Generating Leads Through Search Advertising

Amit Kishore[✉] [iD] and C. Om Prakash

CMR University, Bangalore, India
{amit.17phd,omprakash}@cmr.edu.in

Abstract. A lot of research is available on the effectiveness of search as an advertising channel. Most of these studies tend to treat a click on a search ad as a binary event. All of them study the events leading to the click. This paper goes beyond this to study the post click actions taken by a user subsequent to clicking on a search ad, referring to those actions as post click factors, and testing the factors that have an effect on the final outcome. We use a prescriptive research design employing binary logistic regression analysis. Results indicate that post click factors like the duration of time spent, device used, and recency of visit have a very high positive effect on the final outcome.

Keywords: Forecasting · Marketing · Measurement · AIDA Model · Probabilistic computation · Logistic regression · Lead generation · Post-click factors · Sponsored Search advertising · Sales conversion

1 Introduction

Advertising on search engine is often regarded as a crucial digital advertising platform. Search engine advertising refers to a form of digital advertising when marketers purchase ad placements from a search engine company for particular keywords. The ads are shown next to the results obtained in search (Park & Agarwal, 2018). Important search engine players such as Yahoo!, Bing, and Google have identified paid search advertising as their main source of revenue due to its rapid growth (Domachowski, Griesbaum, & Heuwing, 2016).

Search engines are now an essential component of the decision-making process for consumers. (Goodwin, 2021; Knuth & Masuhr, 2021). Search engines are becoming the go-to source for information on brands among consumers. (Cheng & Anderson, 2021; Deeb, 2021; Statista, 2022).

Search advertising differs significantly from advertising on other channels due to the fact that it is activated when users express their purpose by entering certain keywords or phrases in the search engine while seeking information relevant to them. In that sense, the ads that are hence displayed are closest to satisfy the users' intent. For ex. a user searching for shoes is shown ads related to shoes only, and a user searching for running shoes is shown ads related only to running shoes and not any other type of shoes. Advertisers compete among each other and bid on such keyword to win the right to show their ads before others. The advertiser who wins the bid, gets to show their ad in the first place followed by second and third place winners of such bid. It is worthwhile to note here that search engines show 2 different types of content on its result pages also known as Search Engine Result Pages (SERP) – one ad set paid for by advertisers and another set of results, known as organic results. The organic results are based on the search engine's algorithm. Ads are paid for by advertisers while the organic results cannot be paid for by any advertiser. The advertising costs on different platforms can be calculated either as a share of the sales generated by paid listings or as a predetermined flat amount. Advertisements are displayed at the beginning of a group of results or inside the organic results, and are marked as sponsored (Sharma & Abhishek, 2017). These days, paid search advertising is not only the main source of revenue for search engines but also the most popular method of online lead generation that generates revenue for marketers (Chen, Liu, & Whinston, 2009; Ghose & Yang, 2009; Animesh, Ramachandran, & Viswanathan, 2010).

The biggest advantage of search ads is that they are shown only to people who may have an interest in those ads, as determined by the keywords they have searched for. Therefore, while the ads maybe shown to fewer people than other forms of advertising i.e., display/banner advertising, they are far more likely to result in conversion since almost everyone who is exposed to the ad are interested in those ads. Search advertising, therefore, results in close to zero wastage compared to any other form of advertising. At the same time, search advertising also results in a higher rate of conversion.

In their study, Chalil, Wahana, and Bauman (2020) investigated the impact of search advertisements on the rate of conversion and time taken to sell used cars listed on an internet-based marketplace. Their findings demonstrate that search advertisements have a substantial impact on increasing the likelihood of conversion by an impressive 65.26% and reducing the duration of selling by an average of 3.51 days. A field experiment by Blake, Nosko, and Tadelis (2015) shows that search marketing has no effect on conversion rates for a reputable company such as eBay. The reason being that matching algorithm of the search engine typically places a link to eBay's website at the beginning of the results page. Prior research predominantly employs the Click Through Rate (CTR) and the rate of conversion as key indicators to assess the efficacy of advertising on search engines (Agarwal, Hosanagar, & Smith, 2011; Yang & Ghose, 2010; Rutz & Trusov, 2011). But there is currently no research that examines the impact of post-click factors on conversion.

Purchase conversion can be studied through the purchase funnel or the marketing funnel which in its simplest form is the Awareness-Interest-Desire model (AIDA model). Elias St. Elmo Lewis devised the AIDA model in 1898. It was further explored by Lewis (1903) and Strong (1925) in marketing and advertising literature. The AIDA model is among several models employed to evaluate and analyse the progression of consumers from a state of initial unawareness to the point of making a purchase. As consumers move through the purchase funnel, they move from a state of being unaware to being aware when they enter the funnel followed by a state in which they develop interest in the product, followed by a state in which they develop desire for the product, and culminating in the stage where they take action by buying the product. This last stage is the conversion stage in which the action is determined by the objective of the advertising campaign. An advertising campaign where the objective was to generate leads will end with the consumer filling a form to submit their details as the last action. Lead generation aids client acquisition, conversion, and retention (Chaffey & Ellis-Chadwick, 2019). Similarly, an advertising campaign where the objective was to subscribe to company newsletter, will end with the consumer submitting their email ID as the last action.

Consumers, upon encountering a search ad, have the option to either disregard the ad or click on it. Upon clicking the advertisement, users are directed to the advertiser's website, where they engage in exploring and acquiring information. In most cases, the conversion does not happen in first visit. In case they did not convert on the first visit, they may be exposed to another search ad and visit the advertisers' website again. They may spend some more time browsing and gathering some more information this time as well. In an ad campaign which has ads being served through more than one advertising channels, the consumer may end up on the advertisers' website after clicking on ads on other channels as well i.e., social, display, email, etc. Each visit to the website of the advertise, allows the visitor to acquire additional information pertaining to the product, progressively advancing the consumer through the purchase funnel as they accumulate more knowledge and transition from being aware to becoming interested and then desiring the product, ultimately leading to taking action. Occasionally, customers may choose not to take the intended action on advertisers' websites, even after multiple visits.

Each of these interactions results in different measures of time spent, number of pages visited, page depth inside the website's structure, whether the visit is the first, second, or nth visit, and the time elapsed from the last visit. These factors, when combined, constitute the "depth of interaction" and are all related to actions taken after clicking. It can be argued that the more profound the engagement, the higher the likelihood that individuals are progressing through the funnel and will eventually convert by taking required steps.

This paper studies the direct impact of post-click behaviour on the chance of consumers buying the advertiser's product through conversion. While previous studies have examined pre-click factors such as click-through rate (Goldman & Rao, 2016; Jeziorski & Segal, 2015; Yang & Ghose, 2010), conversion rate (Agarwal et al., 2011; Ghose & Yang, 2009; Rutz, Bucklin, & Sonnier, 2012), and impression (Chan & Park, 2015), in recent years, the importance of click-through rate (CTR) and click-to-conversion rate (CTCVR) has become increasingly evident (Xiao et al., 2022, Xiao et al., 2023). This paper focuses on the influence of post-click behaviour.

This paper attempts to analyse the influence of post-click activities on the final conversion outcome and investigate tactics that advertisers can utilise to improve their conversion rates. Research like this is important for brands to understand actions they must take to increase the chances of conversion for a successful advertising campaign. Some of the factors mentioned above may have a far large effect than others while some may not have any effect. Knowing this can make advertising campaigns more effective and brands can make better use of the advertising budgets. Therefore, the results of this research paper can be used as direct into investment decisions of brands for their advertising campaigns.

2 Review of Literature

Search Engine Marketing (SEM) refers to a kind of internet promotion that utilises well-known search engines. The primary objective is to enhance brand promotion by augmenting the website's visibility (Moran & Hunt, 2009). A Search Engine can assist in promoting a brand's website through two main methods: organic search, also known as Search Engine Optimisation (SEO) and paid or inorganic search advertising (Shih et al., 2013). Paid search, also known as Pay-Per-Click (PPC) or text advertising, functions via a specific mechanism. According to Rutz & Bucklin, 2013, p. 229, optimising a company's listing in search engines' "organic listings," or the search results determined by the algorithms those search engines employ, is the aim of search engine optimisation (SEO). Paid search, on the other hand, enables businesses to purchase a spot on the search engine results page's sponsored or paid listings (SERP). Organic search refers to the search results shown as a result of the search engine's algorithm. Each search engine considers its algorithm the "secret sauce" of its success, as such they would never reveal the actual algorithm that they use to throw up search results. Organic search results are a direct result of the algorithm of the search engine and hence are the main selling item of any search engine. These search results are devoid of any influenced by any outside party and no one can pay their way to a higher ranking in the search results. Brand or companies, in order to gain a long-term sustainable advantage over competition, must strive for a better organic search ranking (Dou et al., 2010). It is crucial to acknowledge that mere inclusion in a search outcome is inadequate, as individuals generally don't venture after the initial result pages. Thus, a website with a high rating is more prone to receiving clicks (Quinton & Khan, 2009).

It is important to observe that unlike advertisements through the pay-per-click method, SEO incurs no expense beyond the amount of time needed for its implementation (Buxton & Walton, 2014, p. 90). To improve search rankings and increase visibility, it is necessary to employ SEO techniques. These techniques include creating a well-organized website structure, developing web content that aligns with commonly searched keywords, and effectively managing both outbound and inbound links to other websites (Xiang & Pan, 2011). Brands or companies wishing to use SEO to gain higher visibility on a search engine can use many SEO techniques (other than mentioned earlier) available to them like adjusting the title tag of the pages on website, meta-tags on all pages, heading tags, and many such tags available on each page, keywords on each page, along with links to and from the website in order to score higher on the search engine's algorithm and, therefore, get a rank that is higher compared to the pages of competing brands as displayed in the SERP (Sen, 2005). SEO techniques are slow to show results, even though the results are more stable once they are start coming. This is in sharp contrast to paid search advertising which shows result when the campaign is active and stop showing results the moment the campaign is stopped. Regrettably, the majority of organisations lack the expertise, training, and perseverance required to successfully implement strategies and patiently await the gradual emergence of results.

Consequently, they prefer to make a payment in order to guarantee immediate website listing and top rankings, provided that they bid high and have a good quality score (Kritzinger & Weideman, 2013, p. 274).

Paid results, unlike organic results, get determined by an auction process. Advertisers participate in this auction by bidding on keywords that users of the search engine are searching for. The auction winners have the privilege of showcasing their text advertisements in descending order based on their bid. An advertiser is not obligated to make a payment just based on winning a bid and having their advertisement broadcast. Conversely, they are only charged when a user actively clicks on the advertisement. Simply said, a search engine generates income solely through user clicks on ads that are shown in response to the user's searched query (Mangold, 2018). This is the reason why paid ads on Google search engine are also called pay-per-clicks advertising. They are known as text ads because Google allows only text to be displayed in these ads. There is no scope to include an image or a graphic or a video in these ads.

The bidding on Google search engine is a unique bidding system which is characterized by two things – 1) it follows the second price bidding model (Edelman et al., 2007), where the highest bidder, instead of paying the amount they bid, only needs to pay an amount that is a fraction of dollar more than the 2nd price bidder in order to outbid them, and 2) bid is not won purely on the basis of the amount bid by each advertiser. Instead, the figure that is taken into account while deciding the winner is a product of amount bid and quality score (Amount X QS). Quality score is a score between 1 & 10 that each advertiser is given by Google based on their past ads and users' reaction to those ads. A higher quality score points to an advertiser whose ads are very effective and users are likely to click on them in the future since in the past they have clicked on them and found the ads useful. A lower quality score, on the other hand, refers to and advertiser whose ads are not very effective and users are less likely to click on them. The objective of the quality score is to ensure 2 things – 1) users get exposed to ads that are likely to be useful to them, and 2) users are likely to click on them ensuring revenue for Google. The quality score, therefore, aims to maximise revenue for the search engine at the same time ensure users get exposed to ads that is useful to them which may influence them to keep coming back to Google again and again. The quality score guarantees that the advertiser with the best revenue potential for the search engine, based on prior performance and ad clicking behaviour, is awarded the top position ad (Rutz & Bucklin, 2013). It can be argued that the quality score also aims to penalize advertisers who serve ineffective ads that are not useful to users who may not click on them or even if they clicked on them, may not find it useful one they reach the advertisers' website. A user who has a bad experience on a search engine will likely not come back depriving it of the revenue it would have earned through clicks.

In 1974, Russell and Mehrabian (1974) proposed the hypothesis of "Stimulus-Organism-Response (S-O-R)". This theory will serve as the foundation for this paper. This theory provides an appropriate framework to investigate the various facets of consumer behaviour (Kim & Lennon, 2013; Rodrıguez-Torrico et al., 2019). This theory is applied in studies examining the impact of online stimuli on consumer behaviour, as noted by Loureiro & Ribeiro (2014) and Kamboj et al. (2018). According to this theory, consumers demonstrate a cognitive and emotional response (organism-O) to specific stimuli (S), resulting in a positive behavioural response (R). While earlier research in this field have studied the effect of stimulus (search ads) on the organism (consumers) who have responded by clicking on the ads (Response) and some have gone even further to take into consideration the conversion (buy) as the response. None of these earlier studies have looked at the behaviour of the consumer after the click on the ads.

Users progress through the search process by entering a search phrase on Google, then selecting an advertisement and ultimately making a purchase on the advertiser's website. The AIDA model offers a structure for comprehending this idea. Elias St. Elmo Lewis first introduced it in 1898 and it was later cited by Lewis (1903) and Strong (1925) in marketing and advertising literature.

2.1 AIDA Model

The AIDA model is a renowned marketing and advertising framework that delineates the four phases of customer decision-making: Attention, Interest, Desire, and Action. It provides a methodical approach to crafting persuasive marketing initiatives and messaging. A consumer, in order to be sufficiently motivated to make the buy, must progress from "being aware" to "being interested" enough to consider the benefits of the product, and from "having the desire" to take advantage of those benefits. As per Lewis, the "Action" phase, therefore, naturally follows the customer's progression through the previous three stages.

Since the majority of consumers do not make a purchase from an advertiser's website during their initial visit, subsequent visits allow the consumer to accumulate increasing amounts of information with each visit. Sahni & Zhang (2024), suggest a constructive role of search advertising where advertising fills significant information gaps by conveying new information that is difficult for the search engines to gather. As a result, the consumer progresses through the various stages of the AIDA model, beginning with awareness and advancing towards action, ultimately resulting in a conversion or sale.

2.2 Post-Click Factors and Depth of Interaction

Figure below depicts a search engine user journey (Fig. 1):

There are several steps involved at **X**, namely, from the moment a user clicks on an advertiser's ad on a third-party website to when the user reaches the advertiser's website and takes an action on that is considered a conversion. Here is a list of actions that a customer may do between clicking on an advertiser's ad on a third-party website and completing a conversion on the advertiser's website:

1. amount of time spent on the advertiser's website in a single session,
2. the quantity of pages viewed on the advertiser's website in a single session,
3. the depth of pages viewed on the marketers' website (as presented),
4. the number of these sessions that occur during the campaign 5. How recent this session is in relation to the previous one.

The device used for ad interaction is another factor that can influence the dependent variable, which in turn affects the device used for the browsing session.

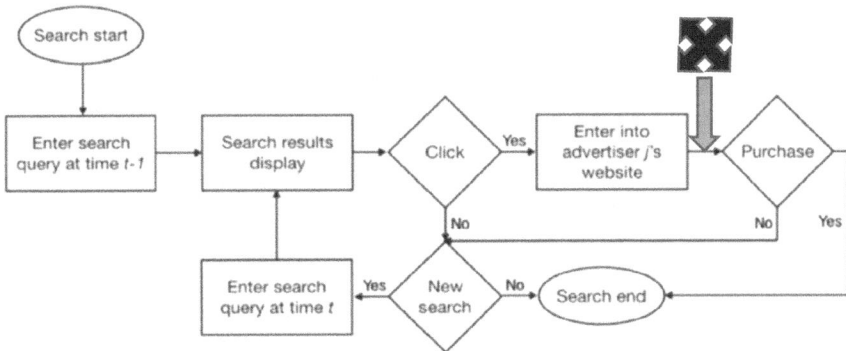

Fig. 1. Customer Search Loop.

Previous researchers have studied the influence of exposure to an ad and the consequent click through rates on consumer purchases or conversions. The current research also illustrates the analysis of the product category, impact of an earlier visit on this visit, the no. of times ad is seen, and the next click that leads to a website visit. Nevertheless, we have insufficient data about the influence of the depth of involvement (any act after the click on ad) on the conversion. This article aims to address a fundamental question: do clicks that result in a greater level of engagement have a higher likelihood of leading to a conversion?

After reviewing the available literature and discussing the topic, the writers have discovered a research gap.

3 The Research Gap

Previous studies in this domain have thoroughly investigated the effects of parameters before the click on the end outcome i.e. purchase or conversion. Nevertheless, there isn't much data on how the post-click parameter—also referred to as the depth of interaction—affects the final outcome.

Heuristics indicate that increasing the values of variables #1, #2, #3, #4, and #5 will result in a larger likelihood of conversion. However, the impact of their combination on the ultimate result remains uncertain.

4 Research Objective

This study has set its objective to analyse how the post click depth of interaction in search advertising channel affects the ultimate outcome, namely whether it results in conversion or no conversion. The final outcome is a binary event, indicated by either a purchase/conversion or the lack of a purchase/conversion.

This research aims to construct a mathematical model capable of predicting if a search advertising campaign will result in a consumer converting or not. This will be accomplished by examining the impact of several parameters associated with the level of engagement on consumer behaviour.

If the final conversion shows a positive correlation with individual variables inside the post click factors, it is probable that these variables, when considered collectively as the depth of interaction, also exhibit a positive correlation. Therefore, the subsequent hypotheses are suggested:

H1 = *average* duration (*time spent*) positively correlates to the final outcome.
H2 = *average no. of pages browsed* positively correlates to the final outcome.
H3 = average *depth of pages* browsed positively correlates to the final outcome.
H4 = *recency of website visit* positively correlates to the final outcome.

In recent years, mobile phones have emerged as the dominant medium for accessing the internet, and it has proven to produce higher likelihood of desired outcomes (such as conversions) in digital advertising campaigns. The authors of this publication put forth a subsequent hypothesis:

H5 = the *device used* on previous visit to the website positively correlates to the final outcome.

5 Research Methodology

5.1 Data and Sampling

The authors are using a prescriptive research design for this study. The authors investigate the post-exposure customer responses to stimuli, specifically search advertising, instead of evaluating their pre-exposure reactions. The data examined in this study relates to the behaviours exhibited by customers following exposure to the commercials. The study measures different characteristics after the initial action of clicking on an advertising.

The data for this research was obtained from a campaign on search engine carried out by a company selling credit cards. The campaign objective was to indue users to click on search ads, navigate to the firm's website, and fill out a form. Consequently, a corporate representative would be prompted to reach out to the consumer in order to facilitate the acquisition of the said credit card. Hence, within the scope of the study, once a form is submitted on the advertiser's website, a sale/conversion is considered as having taken place.

The data analysed spans a 31-day period from January to February 2022, precisely from January 17th to February 16th. This period corresponds to the time when the corporation displayed all of their advertisements.

5.2 Sample Selection

The advertising campaign reached a total of 8,281 distinct consumers over the specified period from January 17, 2022, to February 16, 2022. Out of these individuals, the number of people who were exposed to many advertisements in the campaign and were exposed at least once to a search ad was 1,947. To guarantee the selection of a truly random sample from this aforementioned group, we generated random numbers using the website www.randomiser.org. Samples were selected irrespective of the outcome of the consumer's exposure to a series of advertising, i.e. conversion or non-conversion.

5.3 Sample Size

In Chakraborty & Bhat, 2018, Slovin's formula is used to find a reliable sample size for this study, and it turns out to be anything above 100.

$$\text{Slovin's formula } n = \frac{N}{1 + N * e^2}$$

If the sample size is denoted by "n", and the population size is denoted by "N", and the acceptable margin of error is denoted by "e", the chosen sample is anticipated to yield a confidence level of 90% in our estimations. Based on the computation presented, a minimum sample size of 99 was estimated for the campaign, using a target audience of 15 million. The target audience was defined as Gender – Male/Female, Age – 25 to 44 years, SEC – NCCS A/B, Geography – top 15 cities in India. Using the above definition in Indian Readership Survey, 2019 gave us a total target audience figure of 15 million. Even though the sample size needed was 100, the authors of this paper, having access to data, decided to go for a higher sample size to ensure robustness of the model being created. This research aims to analyse the impact of search ads on 200 customers during the campaign duration.

Although the sample size consists of 200 customers, the data provided by these consumers is extensive due to each consumer undergoing an average of 4 sessions in their consumer journey. Each individual session produces a significant amount of data regarding the independent variables that contribute to the depth of interaction.

5.4 Analysis

IBM Analytics says "Logistic regression is a multivariate technique used to classify a data into groups which is known by estimating the probability of an event occurring or nor not occurring, for example whether customers buy or don't buy, based on a set of the customers characteristics". By utilising a cutoff value, we may assign a specific occurrence to a category based on the probability range of the outcome variable in a logit model, which spans from 0 to 1. The logit function, also known as the logistic function, is defined by the following expression:

$$Logit(y) = \frac{1}{1 + \exp(-y)}$$

$$\text{where } y = \beta_0 + \beta_1 * x_1 + \ldots\cdots + \beta_n * x_p$$

Logit(y) in this case denotes the response or dependent variable, whereas the independent variables are represented by $x_1, x_2, \ldots, x_p$ and the co-efficient for each variable is represented by $\beta_1, \beta_2\ldots, \beta_p$ with β_0 being the independent variable for the logistic regression equation. The conditional probability of each observation is calculated. The value is combined to ascertain the expected probability for each observation. For classification that is binary in nature, probabilities below the threshold of 0.5 will be classified as 0, while probabilities over 0.5 will be classified as 1. The efficacy of the calculated model is determined by applying it to a separate data set to ascertain its level of accuracy.

Table 1. Descriptive statistics for the scaled variables.

Descriptive Statistics

	N	Min	Max	Mean	Std. Deviation
Search	161	1	3	1.93	0.771
Dur_Search	161	5	993	409.40	262.229
Pages_Search	161	1	7	2.96	1.470
Depth_Search	161	1	5	3.01	1.410
Recent_Search	161	1	14	5.42	3.059

6 Data Analysis and Interpretation

The descriptive statistics of the scaled variables are represented in Table 1. Search refers to the frequency at which a user interacted with Search Ads by clicking on it. The maximum value in this case is three times of the average, which is approximately 1.93 and close to 2. The smallest duration is 5 s, while the greatest duration is 993 s. Each visit has an average duration 409 s, which is nearly 7 min. The max number of pages is 5, and the maximum depth for every page is also 5, average depth being 3, which is quite substantial. The maximum time spent during recent visit is 14 s which comes to an average time spent of 5.42 s.

Table 2. Frequency of the dependent variable – no. of users purchased

Page Event

		Frequency	Percent	Valid Percent	Cumulative Percent
Valid	No Purchase	91	56.5	56.5	56.5
	Purchase	70	43.5	43.5	100.0
	Total	161	100.0	100.0	

Table 2 above displays the frequency for the dependent variable and we see that there are 91 visitors out of a total of 161 who did not purchase and that is 56.5%. Only 43.5% of the visitors i.e., 70 out of 161 purchased the product.

The above Table 4 represents the categorization as determined by the resulting model obtained from the binary logistic regression. The success rate is 69%, indicating that the model has been able to accurately classify 69% of the original events. The selected threshold was set at 0.5. The model has higher accuracy in classifying those who have not made a purchase (84.8%) compared to those who have made a purchase (47.8%).

The regression coefficients are displayed in the above Table 5. This helps us in understanding the role of the significant variables in forecasting the correct set in the target variable, i.e., Page Event. These variables include Search times, Duration of the Search, recent visit time spent, and the device last used. The significance values of these

Table 3. Model Summary of binary logistic regression

Summary of the Model

Step	−2 Log likelihood	Cox & Snell R Square	Nagelkerke R Square
1	195.540[a]	0.140	0.188

a. Estimation termination at 4th iteration because estimates differ by less than 0.001

Table 4. Classification table

Classification Table[a]

	Observed		Predicted		Percentage Correct
			Page Event		
			No Purchase	Purchase	
Step 1	Page Event	No Purchase	78	14	84.8
		Purchase	36	33	47.8
	Total Percentage				68.9

a. 0.500 is cut-off value

Table 5. Variables table with their parameter values

Equation variables

		B	S.E.	Wald	df	Sig.	Exp(B)
Step 1[a]	Search	0.551	0.229	5.876	1	0.015	1.736
	Dur_Searched	0.002	0.001	5.778	1	0.016	1.002
	Pages_Searched	0.046	0.118	0.156	1	0.693	1.047
	Depth_Searched	−0.118	0.126	0.904	1	0.343	0.887
	Recent_Searched	−0.135	0.063	4.791	1	0.029	0.874
	Device_LastVisited	0.883	0.352	6.261	1	0.012	2.423
	Constant	−1.544	0.809	3.645	1	0.056	0.213

a. Variable(s) entered in step 1

coefficients are all less than 0.05. The number of pages and length of "Search", which refer to the number of times visited and pages visited, don't seem to exert an impact on determining the target variable group. In Table 3, 0.140 is the Cox and Snell R2 value, while 0.188 is the Nagelkerke R2 value. These values indicate a rather excellent fit for the model, and they have statistical significance when compared to the base model that does not include any predictors.

The coefficient (B) value, on the other hand, indicates that all variables, except Pages_Searched, have a positive correlation with the dependent variable page event. This is evident from the fact that the coefficient value is above 1 for these variables. Wald statistics indicate that the Device_Used for the last visit (6.316) is the variable with the highest importance, next is the depth of pages visited (1.823), the frequency of a search result being clicked (1.346), the Recent_Searched (1.192), and finally the Duration_search (1.003).

As a result, the model looks thus:

Y = − 1.541 + 0.551*Search + 0.002*Duration of Search − 0.12 * Depth of Search − 0.14* Recency of Search + 0.05* Pages Searched + 0.883 * Device in Last visit.

$P = e^Y / 1 + e^Y$ is the logistic equation which gives the probability of a particular visitor purchasing the product. This model will suggest what is the probability of a visitor going towards buying based on the Search. The higher the probability (P), with a maximum value of 1, the greater the likelihood of the visitor making a purchase. In other words, a page event with a value of one (1) indicates that the visitor will buy the product.

7 Discussion and Conclusion

This study presents a model that estimates the probability of consumers converting after being exposed to an ad through more than one digital media channels, which includes at least one ad in search medium. Table 5 demonstrates that certain variables within the depth of interaction have a statistically significant impact (sig < 0.05) on the prediction of the target variable.

Recency of Display = 0.029 < 0.05
Duration of visit = 0.016 < 0.05
Device last used = 0.012 < 0.05

Based on the above data, hypothesis H1, H4 & H5 are accepted.

The data also suggests that the following variables have significance value even though they are outside the acceptance limit:

Depth of Pages = 0.694 (sig value > 0.05)
No. of *Pages* = 0.198 (sig value > 0.05)

Based on the above, we must reject the hypotheses H2 & H3.

The Cox & Snell pseudo R2 score of 36.6% and the Nagelkerke pseudo R2 value of 48.8% suggest that the model fits well. The Wald test indicates that the variables that have the most significant influence on the final result, listed in descending order, are Device utilised, Duration of visit in a session, Depth of pages visited, and Recency of visit.

Exp (B) coefficients reveal that Duration, Depth, Page, and Device are the variables with coefficients exceeding 1, indicating that an increase in these values is associated with a higher chance of a positive result. Depth & Recency of visit have Exp (B) value below 1, indicating that any increase in value would lead to a drop in the likelihood of achieving the end result, i.e. conversion. The negative B value of each of these variables supports this conclusion.

Prior research (Chatterjee et al. (2003)), could not find a link between each visit or session. Nevertheless, the authors of this paper effectively show the importance of the timing of a visit. More precisely, when a visitor clicks a search ad at a later stage in their journey, it does have a big impact on the end result. The authors of this paper would like to highlight the substantial influence of the Device_Used in forecasting the outcome.

8 Contribution to Theory

There is multiple value addition by this paper to theory – significant among them is the ability to consider a user's behaviour after they have clicked on as ad. Very few researchers have attempted to study this. Therefore, a significant outcome of this research is the link between exposure to an ad (determined by an ad click) and the following response from the consumer resulting in a conversion. Without this linkage, any click on a search ad is assumed as a binary event, but this research paper is able to prove that this event takes a value that ranges from 0 to 1, which depends on how deep the interaction was after clicking on the ad.

9 Managerial Implications

The most significant learning from this paper, for managers, is criticality of search advertising for an ad campaign's success. (based on the final outcome and not just the exposure). Managers will be able to fine tune their budgets for search campaign based on the depth of interaction, instead of only the ad click (binary value 0/1).

This paper establishes that the parameter that is most significant to the final outcome is device used to reach advertiser's website. Hence, managers must make sure their website is responsive and fits the device on which the ad is running.

This research also establishes the role of "Depth of pages visited", apart from the "Duration of the visit" and "Recency of visit". This should lead to campaigns having good spacing between each ad exposure. Also, the content on the website must make the users dwell more and explore pages that are deeper in the hierarchy instead of staying at top-level pages (i.e., homepage). This means managers need to think beyond just how a campaign is created but also how the advertiser website must be. Some simple changes to the ad copy and website content, with the device used in mind, can surely lead to a better return on investment by delivering more conversions.

10 Future Research

One of the most obvious limitations of this research is that it does not take into account the gender, age, etc. of the consumer into account. Introducing such consumer profile data can enhance this research. It may be worth seeing how consumer respond differently based on their demographic and other parameters.

This research paper does not take into account the effect of competition on the consumers' behaviour on the ad and the website. While within a single campaign, in real life, consumers get exposed to many different channels, this study looks at search in isolation. The dependent variable (conversion) is actually influenced by more than just one medium (search). It would be interesting to see what happens to the convergence when more than one channel is used.

Disclosure of Interests. The authors have no competing interests to declare that are relevant to the content of this article.

References

1. Animesh, A., Ramachandran, V., Viswanathan, S.: Research note—quality uncertainty and the performance of online sponsored search markets: an empirical investigation. Inf. Syst. Res. **21**(1), 190–201 (2010)
2. Blake, T., Nosko, C., Tadelis, S.: Consumer heterogeneity and paid search effectiveness: a large-scale field experiment. Econometrica **83**(1), 155–174 (2015)
3. Buxton, M., Walton, N.: The Internet as a small business e-commerce ecosystem. In: Lacka, E., Chan, H., Yip, N. (eds.) E-commerce platform acceptance: Suppliers, retailers, and consumers, pp. 79–100 (2014). https://doi.org/10.1007/978-3-319-06121-4_5
4. Chaffey, D., Ellis-Chadwick, F.: Digital Marketing. Pearson (2019)
5. Chakraborty, U., Bhat, S.: The effects of credible online reviews on brand equity dimensions and its consequence on consumer behavior. J. Promot. Manag. **24**(1), 57–82 (2018)
6. Chalil, T.M., Dahana, W.D., Baumann, C.: How do search ads induce and accelerate conversion? The moderating role of transaction experience and organizational type. J. Bus. Res. **116**, 324–336 (2020)
7. Chan, T.Y., Park, Y.H.: Consumer search activities and the value of ad positions in sponsored search advertising. Mark. Sci. **34**(4), 606–623 (2015)
8. Chatterjee, P., Hoffman, D., Novak, T.: Modeling the clickstream Implications for Web-Based Advertising Efforts. Mark. Sci. **22**(4), 520–541 (2003)
9. Chen, J., Liu, D., Whinston, A.B.: Auctioning keywords in online search. J. Mark. **73**(4), 125–141 (2009).
10. Cheng, M., Anderson, C.: Search engine consumer journeys: exploring and segmenting click-through behaviors. Cornell Hospitality Q. **62**(2), 198–214 (2021)
11. Deeb, G.: All search engine traffic is not created equal (2021). https://www.forbes.com/sites/georgedeeb/2021/09/02/all-search-engine-traffic-is-not-created-equal/?sh=74f4d7d5c561
12. Domachowski, A., Griesbaum, J., Heuwing, B.: Perception and effectiveness of search advertising on smartphones. In: Proceedings of the 79th ASIS&T annual meeting: Creating knowledge, enhancing lives through information & technology, pp. 74–84. American Society for Information Science (2016)
13. Lewis, E.: "Catch-Line and Argument 3(11), e2." The Book-Keeper **15**, 124 (1903)
14. Faulds, D.J., Mangold, W.G., Raju, P.S., Valsalan, S.: The mobile shopping revolution: redefining the consumer decision process. Bus. Horiz. **61**(2), 323–338 (2018)
15. Ghose, A., Yang, S.: An empirical analysis of search engine advertising: sponsored search in electronic markets. Manage. Sci. **55**(10), 1605–1622 (2009)
16. Goldman, M., Rao, J.: Experiments as instruments: heterogeneous position effects in sponsored search auctions. EAI Endorsed Transactions on Serious Games (2016)

17. Goodwin, D.: 71 Mind-blowing search engine optimization stats (2021). www.searchenginejournal.com/seo-guide/seo-statistics/#close
18. Jeziorski, P., Segal, I.: What makes them click: empirical analysis of consumer demand for search advertising. Am. Econ. J. Microeconomics **7**(3), 24–53 (2015)
19. Kamboj, S., Sarmah, B., Gupta, S., Dwivedi, Y.: Examining branding co-creation in brand communities on social media: applying the paradigm of Stimulus-Organism-Response. Int. J. Inform. Manag. **39**, 169–185 (2018). https://doi.org/10.1016/j.ijinfomgt.2017.12.001
20. Kim, J., Lennon, S.J.: Effects of reputation and website quality on online consumers' emotion, perceived risk and purchase intention: Based on the stimulus-organism-response model. J. Res. Interact. Mark. **7**(1), 33–56 (2013)
21. Knuth, I., Masuhr, J.: Decision drivers for search engine usage–the mediating and moderating role of lock-in effects. Int. J. Bus. Manag. **16**(10), 86 (2021). https://doi.org/10.5539/ijbm.v16n10p86
22. Kritzinger, W.T., Weideman, M.: Search engine optimization and pay-per-click marketing strategies. J. Organ. Comput. Electron. Commer. **23**(3), 273–286 (2013)
23. Moran, M., Hunt, B.: Search Engine Marketing, Inc.: Driving Search Traffic to your Company's Web site. IBM Press (2009)
24. Pan, B., Xiang, Z., Law, R., Fesenmaier, D.R.: The dynamics of search engine marketing for tourist destinations. J. Travel Res. **50**(4), 365–377 (2011)
25. Park, C.H., Agarwal, M.K.: The order effect of advertisers on consumer search behavior in sponsored search markets. J. Bus. Res. **85**, 24–33 (2018)
26. Rodrıguez-Torrico, P., San-Martın, S., San Jose-Cabezudo, R.: What drives M- shoppers to continue using mobile devices to buy? J. Mark. Theory Pract. **27**(1), 83–102 (2019). https://doi.org/10.1080/10696679.2018.1534211
27. Russell, J.A., Mehrabian, A.: Distinguishing anger and anxiety in terms of emotional response factors. J. Consult. Clin. Psychol. **42**(1), 79 (1974)
28. Rutz, O.J., Trusov, M.: Zooming in on paid search ads—a consumer-level model calibrated on aggregated data. Mark. Sci. **30**(5), 789–800 (2011)
29. Rutz, O.J., Bucklin, R.E., Sonnier, G.P.: A latent instrumental variables approach to modeling keyword conversion in paid search advertising. J. Mark. Res. **49**(3), 306–319 (2012)
30. Sahni, N.S., Zhang, C.: Are consumers averse to sponsored messages? The role of search advertising in information discovery. Quant. Mark. Econ. **22**, 63–114 (2024)
31. Sharma, S., Abhishek, V.: Effect of sponsored listings on online marketplaces: the role of information asymmetry (2017)
32. Statista: U.S. Search engines: number of core searches (2022). https://www.statista.com/statistics/265796/us-search-engines-ranked-by-number-of-core-searches/
33. Strong, E.: The Psychology of Selling and Advertising. McGraw-Hill, New York (1925)
34. Xiao, Y., He, W.K., Zhu, Y., Zhu, J.: A click-through rate model of e-commerce based on user interest and temporal behaviour. Expert Syst. Appl. **207**, 117896 (2022)
35. Xiao, Y., Zhu, Y., He, W.K., Huang, M.: Influence prediction model for marketing campaigns on e-commerce platforms. Expert Syst. Appl. **211**, 118575 (2023)
36. Yang, S., Ghose, A.: Analyzing the relationship between organic and sponsored search advertising: positive, negative, or zero interdependence? Mark. Sci. **29**(4), 602–623 (2010)

mCLIP: Multimodal Approach to Classify Memes

M Kaab Bin Shahid[1,2](✉) , Hamid Husain[1,2] , and Hira Javed[1,2]

[1] Department of Computer Engineering, Zakir Husain College of Engineering and Technology, Aligarh, India
m.kaabbinshahid@gmail.com, gk6273@myamu.ac.in, hira.javed@zhcet.ac.in
[2] Aligarh Muslim University, Aligarh, India

Abstract. The exponential rise of internet memes, often humorous images with embedded text, across major social media platforms like Facebook, Instagram, and X (formerly Twitter) has sparked significant attention in recent years. This phenomenon, while entertaining for many, has brought to light a daunting challenge, the high occurrence of hate speech on these platforms. Addressing this challenge has become a communal responsibility, placing considerable pressure on social media companies to mitigate the spread of harmful content. In response to this escalating concern, our research introduces an innovative solution named "mCLIP," which is a variation of CLIP. mCLIP employs a multimodal approach to effectively classify memes based on their levels of offensiveness and positivity. This paper discusses the advancement and execution of mCLIP designed for multilevel and binary classification of memes by focusing on evaluating its effectiveness in distinguishing between harmless humor and potentially harmful content. By utilizing advanced techniques in multimodal analysis, mCLIP contributes to the ongoing conversation about creating a safer and more responsible digital environment. We have performed our task on SemEval 2020 task 8 dataset [20] and our results outperformed SOTA models. Furthermore, the outcomes highlight the importance of taking proactive steps to deal with the issues raised by the changing nature of internet memes on social media.

Keywords: Multimodal · CLIP · Meme

1 Introduction

In the domain of sentiment analysis within social media, text-based approaches have achieved remarkable accuracy, often ranging up to 95%. While this success has paved the way for effective understanding of textual content, the classification of memes presents a unique challenge. At first glance, it might seem straightforward, with options such as Optical Character Recognition (OCR) [18] for text encoding or Convolutional Neural Network (CNN) models for image-based classification. However, the complexity arises from the fact of relying solely on one modality, be it text or image, to accurately classify memes. Take, for instance Fig. 1, the meme that reads, "look how many people love you."

If a model exclusively focuses on text, it might categorize the meme as non-offensive due to the seemingly pleasant text. Conversely, a model solely based on image analysis may perceive the meme as non-informative and non-offensive as solely the image is not conveying anything.

Fig. 1. Meme

The true challenge lies in integrating both modalities simultaneously, as a combined interpretation of text and image reveals a different classification and categories the meme as potentially offensive or hateful, contrary to the initial assessment based on individual modalities. This complexity emphasizes the need to consider both textual and visual elements in this task. Therefore, this research aims to address the challenges inherent in meme classification by using a multimodal approach and not solely depending upon only one modality. As memes are sarcastic and contains various logical and humorous features, therefore the right feature extraction technique must be applied as understanding sarcasm is seen as a task that present models cannot perform accurately. Our contribution are as follows:

- A novel approach that uses only robust CLIP encoders for classifying memes
- Multilevel classification model for memes that outperforms the state-of-the-art models
- Competitive binary meme classification model

2 Related Work

Researchers have proposed a spectrum of methodologies aimed at classifying and detecting the nature of memes. While some studies concerned on only binary classification taking into account whether a meme is offensive or not, some studies worked on multilevel classification.

SemEval-2020 [1] focuses on the computational analysis of Internet memes in multiple classes, emphasizing their multimodal nature. The Memotion analysis task released

7,000 annotated memes and include subtasks for sentiment analysis, overall emotion classification, and intensity classification.

While certain memes aim for direct humor, others opt for a sarcastic take on everyday life occurrences. In this comprehensive meme emotion recognition framework, they employ a baseline model, binary classifier for text content analysis, utilizing word embeddings fed into a CNN-LSTM architecture. Simultaneously, image content analysis relies on pre-trained models like VGG-16, ResNet-50, and AlexNet to extract features. With the strengths of both modalities, producing predictions for sarcasm, humor, offense, and motivation. The baseline model addressed the efficiency of individual content modalities, text and image, and their combination. Notable models [2–6], showcased diverse methodologies, including ensemble learning, transfer learning, and modality ensembles.

[7] proposes a potential to reduce the amount of manual labor at social media companies if multimodal models are used for the task. They have used the dataset from the Hateful Meme Challenge provided by Meta. Multiple types of models and various methods of multimodal fusion to obtain classifications on memes were tested. In the implementation three baseline models from the original challenge were used [8], Basic Late Fusion, ConcatBERT, and VisualBERT. Additional they have also explored different multimodal models like CLIP [9] and BridgeTower [10] which uses different fusion methods. In establishing a baseline, they constructed two unimodal models. The first utilized a pre-trained BERT model (Bidirectional Encoder Representations from Transformers) from HuggingFace to classify the textual content of memes. Simultaneously, a second model employed a pre-trained ResNet 152 model to classify the images within the memes. Both of these models were fine-tuned using the Hateful Memes dataset. To enhance the baseline, a basic late fusion model was developed. This involved combining the normalized logits from both unimodal models by summing them, thereby creating a unified set of logits for subsequent classification. The analysis encompassed baseline unimodal, late fusion model and early fusion models such as ViLT, CLIP, ConcatBERT, and VisualBERT. The results indicated that early fusion models that take multimodality into account consistently outperformed basic unimodal and late fusion counterparts. Among them, CLIP demonstrated the highest performance. Also, BridgeTower and ViLT showed competitive performance over other models which clearly shows the importance of taking image-text relationship in recognizing hateful memes. While early fusion models showcased advancements over unimodal and late fusion models, the study identified the impact of fusion methods on model effectiveness. Models with detailed interaction mechanisms (CLIP, BridgeTower, ViLT) surpassed those relying on basic concatenation (ConcatBERT).

In SemiMemes [11] the input (image, text) is generated by CLIP feature vectors (Fimage, Ftext). Stage 1 involves unsupervised pre-training with Cross Modality Auto Encoder (CROM-AE), predicting features across modalities using unlabeled data. Stage 2, supervised fine-tuning, freezes CROM-AE encoders, extracts new representations, and fuses them with original CLIP features for classification in the Raw and Cooked Features Classification Model. SemiMemes outperforms CMML-CLIP [12] and TIB-VA [13] in all label settings for the MAMI benchmark. In the 5% label setting, SemiMemes achieves a test score of 0.693, surpassing CMML-CLIP and TIB-VA by 0.034 and 0.039, respectively. Performance gaps decrease with more labelled samples. In the 30% labelled setting, SemiMemes performs competitively. It achieves slightly higher scores less than 1% than Hate-CLIPper [14]. Despite similar performance, SemiMemes is more efficient with significantly fewer parameters (3.1 million) compared to Hate-CLIPper (1.5 billion), thereby saving training resources while maintaining good performance. Moreover, [8] introduces a novel multimodal classification challenge centered on detection of hateful content in memes, it introduces "benign confounders," alternative images or captions that flip the label from hateful to not-hateful and vice versa. This strategy counters the model exploitation of unimodal priors and emphasizes the need for sophisticated multimodal reasoning. Solving this challenge requires AI models capable of understanding between original memes and their confounders. In the broader context, the challenge aligns with the increasing role of machine learning in addressing societal issues, such as hate speech detection at the internet scale, emphasizing its real-world application. The dataset described in the paper [8] serves as a set of competitiveness for finetuning and testing models with large multimodal capabilities pre trained through self-supervised manner. For mitigating visual predisposition and adhere to licensing requirements, it reconstructed original memes from ground using a custom mechanism. Moreover, they utilized arbitrator annotators who gave nearly 27 min for each meme to get annotated to the final format. Their work assesses models falling into three categories: unimodal models, multimodal models pre-trained with unimodal pretraining, and multimodal models pre-trained with a multimodal objective. For the imaging modality, it uses standard ResNet-152 convolutional features (Image-Grid) and characteristics from the fc6 layer of Faster-RCNN with ResNeXt-152 as its backbone (Image Region). In unimodal model for text is BERT (Text BERT). Furthermore, it compares simple fusion methods like taking mean of output scores of ResNet-152 and BERT or Late Fusion like Concat-BERT (BERT features concatenated with ResNet-152 over which an multilayer perceptron is trained) with more sophisticated multimodal methods, including supervised multimodal BiTransformers (MMBT-Region and MMBT-Grid) and types of pretrained Visual BERT and ViLBERT. The paper also takes into account hyperparameter tuning involving grid search. The result reveals that the vision classifier was defeated by a slight margin by text classifier counterpart. text-only classifier slightly outperforms the vision-only counterpart. Notably, multimodal models demonstrate improved performance, with early fusion models (ViLBERT [23], MMBT and VisualBERT [21]) generally outperforms concatenation and late fusion approaches. The results also shows that by using pretrained unimodal with fusion and pretrained multimodal the performance difference comes very low, it may seem that multimodal approach surpasses the unimodal one but as the annotators' accuracy as listed in the respective paper is nearly 85% and the highest achieved accuracy by multimodal pretrained model comes out to be nearly 75% (using pretrained VilBERT and VisualBERT), therefore the findings of the paper suggests that there is a high scope for improvement in the multimodal approach.

Other works like [16] uses visual and textual feature extraction for classifying the memes, [17] proposed 2 models with text only classification using OCR of meme and a multimodal approach using meta hateful meme dataset, [18] classified memes by OCR tesseract with combination with Naïve Bayes algorithm and [19] proposes binary classification model using SwinT and BiLSTM approach.

In our research paper, we present a novel approach to hate speech detection in multimodal memes, departing from certain strategies employed by existing models such as Hate-CLIPper and the Visual BERT COCO [21] variant. Unlike Hate-CLIPper, which relies on the Feature Interaction Matrix (FIM) for similarity projection between images and corresponding text, our dataset comprises memes with 100% similar texts corresponding to the images. As a result, we have opted not to use the FIM of CLIP in our model. Instead, we leverage the robust pre-trained encoders of the CLIP model, due to the inherent alignment between text and image in our dataset. Moreover, in the HMC Challenge Next Move paper [15], the authors fine-tuned the Visual BERT COCO model by altering hyperparameters, yielding two distinct models that achieved high test accuracy. However, insights from [7] emphasize the necessity of multimodal models for meme classification, with CLIP emerging as a promising candidate according to various studies. Contrary to the approach of finetuning CLIP, as seen in Hate-CLIPper, our research contends that this may not be an optimal strategy. [7] also reveals that finetuning CLIP led to lower accuracy compared to the base CLIP model, indicating potential challenges in adapting a model trained on a diverse range of image-text pairs to the specific nuances of memes. As Memes are inherently satirical where often feature text diverges from the literal interpretation of the accompanying image and focuses more on Optical Character Recognition (OCR) of the image. To mitigate the potential confusion arising from finetuning, our model avoids this step and exclusively utilizes the robust text and visual encoders of CLIP. By steering clear of finetuning, we aim to maintain the integrity of CLIP's pre-trained capabilities on a vast corpus of image-text pairs because the unique challenge posed by memes, where textual elements may not directly correspond to visual content. Our methodology underscores the importance of preserving the nuanced understanding inherent in pre-trained models, especially when applied to the complex and often satirical context of multimodal memes.

3 Methodology

The dataset [20] utilized for this research comes from SemEval 2020 Task 8, publicly accessible on the SemEval website and other downloadable datasets repositories. Comprising a total of 6992 images, each image is accompanied by its corresponding text, extracted through Optical Character Recognition (OCR), and catalogued in an Excel file. This rich dataset encompasses 52 distinct and globally recognized categories, featuring well-known entities such as Hillary, Trump, Minions, Baby Godfather, and more.

The dataset also encapsulates information on five key categories: humor, sarcasm, offensiveness, motivation, and overall sentiment. The overarching sentiment category is further categorized into three classes: positive, neutral, and negative. In order to facilitate model training and evaluation, the dataset has been divided into an 80–20 ratio for training and testing as mentioned in Table 1.

Table 1. Dataset distribution

Dataset	6992
Training	5693
Testing	1399

3.1 Sentiment Classification Logic

The classification of memes into positive, negative, and neutral categories is anchored by the following points and does not depend upon the level of offensiveness.

- Meme conveying sarcastic or logical information of high degree is categorized as positive.
- Meme conveying sarcastic or logical information of medium to low degree is categorized as neutral.
- Meme conveying sarcastic or logical information of very low degree is categorized as negative.

This approach to sentiment classification prioritizes the overall information content of the meme, recognizing that the conveyed message, irrespective of its nature, forms the basis for assigning a positive, negative, or neutral sentiment label.

3.2 Preprocessing

The preprocessing phase is crucial for maintaining the accuracy and entirety of the dataset. Given the occasional absence of data, especially image files corresponding to text and image names listed in the Excel file. Each image name from the CSV file is cross-referenced with the contents of the image folder. Valid names are appended with their respective paths, thereby rectifying any instance of missing data. Any remaining gaps are addressed through data elimination. Our classification approach involves two distinct strategies, multilevel classification and binary classification. In the case of multilevel classification, where labels include positive, neutral, and negative, label encoding is applied. Similarly, for binary classification focusing on offensive or non-offensive labels, label encoding is employed.

Given the constraints of the text input length, as CLIP take a maximum of 77 tokens, hence truncation of textual data needs to be done, derived from Optical Character Recognition (OCR) of the images.

3.3 Technique Used: Contrastive Language-Image Pre-training

CLIP [9] is leading the latest advancements in artificial intelligence where it demonstrates its potential as a multimodal model. Developed by OpenAI, CLIP has the ability to comprehend both textual and visual information thereby cutting down the traditional barriers between language and images. What sets CLIP apart is its unique pre-training process, which involves learning from a vast array of internet images and their associated textual descriptions in a contrastive learning format. This learning approach as shown in Fig. 2 enables CLIP to understand the intricate relationships between different modalities, allowing it to connect words with images in a nuanced and contextually rich manner. Unlike traditional models that focus solely on either text or images, CLIP handles both domains. This versatility leads to a wide range of applications, including image classification, object recognition, natural language understanding, and textual entailment. Moreover, due to zero-shot prediction capability of CLIP, it makes it very powerful and appealing for researchers to use it in a classification task.

CLIP employs separate transformer models to encode both textual and visual features and then the transformer's attention mechanism allows it to capture intricate relationships within the input data, whether it be in the form of words in a sentence or features in an image.

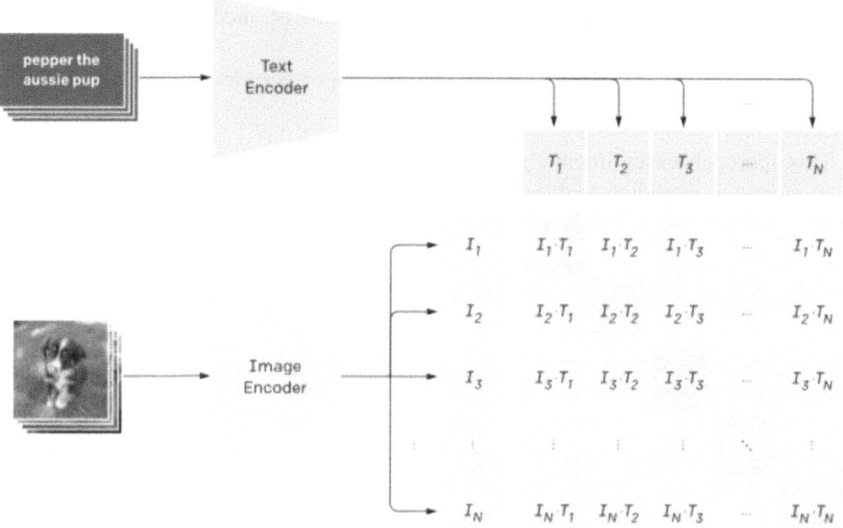

Fig. 2. CLIP Architecture [9]

In the multimodal space these encoded representations from text and images are brought together. During training, CLIP utilizes contrastive learning approach. This involves positioning similar image and text pairs close to each other in the feature space while pushing dissimilar pairs farther apart. By doing so, CLIP learns to understand the semantic connections between the two modalities. The loss calculated based on this contrastive learning is then backpropagated through the neural network, updating the parameters of both the text and image encoders. This iterative process refines the representations learned by the encoders, making them increasingly robust and adept at capturing the nuances of both textual and visual information.

3.4 Justification for CLIP Usage

In our research project, the crucial interaction between images and text calls for a model that deeply understands the complex relationships within image-text pairs. Given this need, choosing CLIP is both logical and strategic. CLIP is specifically designed for multimodal tasks like these and excels at comprehending the relationship between images and text. The project requires strong encoders trained through contrastive learning, capable of effectively encoding both textual and visual information and should not be using late fusion or concatenation methodology. CLIP's encoders, which have been extensively pre-trained, have the nuanced understanding needed to effectively encode different image and text pairs. Our method involves using these powerful encoders to encode the images and their associated text in our dataset. This process produces embeddings that contain the semantic essence of the memes, providing valuable information. These encoded embeddings, derived from CLIP's adept comprehension of image-text relationships and forms the basis for our meme classification tasks.

We chose to use CLIP in our research because our multimodal project requires a model that can handle the complex interaction between images and text. CLIP's robust encoders and contrastive learning approach are well-suited to our goals for effective meme classification.

The role of CLIP in our model whether it is binary classification or multilevel classification of memes is for generating image and text encodings which works in neural network as an input.

3.5 Our Models

3.5.1 Multilevel Classification Model Architecture

In our proposed multilevel classification model as in Fig. 3, the goal is to categorize memes into positive, neutral, or negative sentiments, providing an overarching understanding of their emotional tone. The input layer of mCLIP accepts text embeddings and image embeddings from CLIP. Given that each of these embeddings is of 512 dimensions, the input layer is configured with 1024 input dimensions (512 for text and 512 for image), housing 128 units.

Following the input layer, the model incorporates a sequence of six hidden layers. These hidden layers are structured with varying number of neurons: 512, 1024, 1024, 512, 64, and 32. Dropout layers are set on the first three hidden layers with dropout rates of 0.15, 0.20, and 0.25, respectively. The introduction of dropout is a key aspect aimed at mitigating potential overfitting issues. Dropout, a regularization technique, works by randomly removing units during training. This prevents the model from relying too much on specific neurons, improving its ability to generalize. This helps the model stay strong and effective when trained with various meme data. After the hidden layers, the model ends with the output layer, set up with three units. This setup matches the model's goal of categorizing memes into three clear classes: positive, neutral, and negative sentiments. For all layers except the output layer, Rectified Linear Unit (ReLU) activation is employed. ReLU introduces non-linearity, facilitating the model's ability to capture complex patterns in the data. In the output layer, for multilevel classification task, the softmax activation function is used. This is particularly beneficial in the context of our three-class classification, where each meme is assigned a probability distribution across the positive, neutral, and negative classes.

We use sparse categorical cross-entropy as the loss function for training, which is suitable when classes are represented as integers. Since we've label encoded positive, neutral, and negative classes, this loss function is appropriate for assessing the difference between predicted and actual class distributions. The determination of the count of layers and units in the hidden layers is predominantly an experimental process governed through practical considerations. Our approach involved a systematic exploration, testing the model's performance across a spectrum of configurations, including 3, 4, 5, and 6 layers. Within each layer, the number of neuron units spanned a range from 32 to 4096. Through extensive experimentation, we aimed to find the setup that provided the best accuracies and strong model performance. After numerous trials, the proposed architecture, with its specific layer and unit combination, proved to be the most effective. Notably, this configuration showed superior accuracy and skill in handling the nuanced task of meme classification.

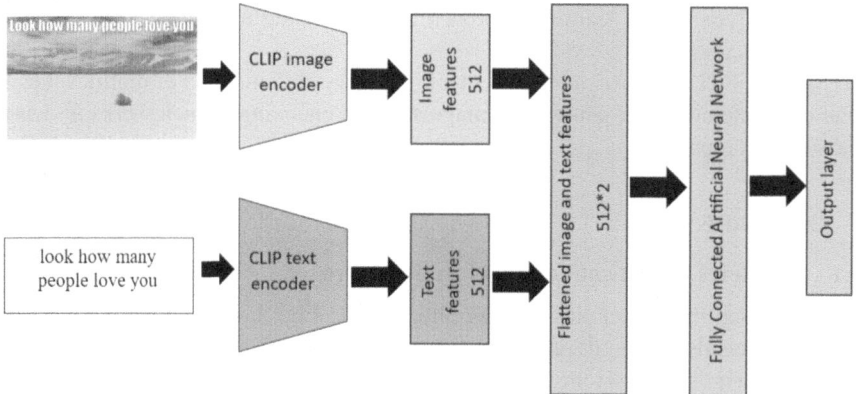

Fig. 3. Multilevel model architecture, fully connected layers.

We base our decision-making on performance metrics from thorough testing, ensuring that the selected architecture fits well with the dataset's complexities and the classification task's intricacies.

3.5.2 Binary Classification Model Architecture

For binary classification architecture shown in Fig. 4, aimed at distinguishing offensive from non-offensive content, we've utilized a multilevel model architecture with careful adjustments to fit the binary classification's requirements. In this version, the output layer is customized to meet the needs of binary classification. It consists of only one unit due to the binary nature of the task as to determine whether a meme is offensive or not. To facilitate training for this binary classification, the model utilizes binary cross-entropy as the loss function. This choice of loss function is well-suited for scenarios where the task involves assigning instances to one of two exclusive classes.

Label encoding is utilized for encoding predictable output class. This encoding simplifies the representation of the target classes, assigning the label "1" to signify offensive content and "0" for non-offensive content. The activation function at the output layer is adjusted to sigmoid activation. Sigmoid activation transforms the raw output into a probability score between 0 and 1, effectively capturing the model's confidence in classifying a meme as offensive. The training regimen for the mCLIP model spans a duration of 10 epochs. This implies that the model undergoes iterative learning and refinement over the entire dataset for 10 complete passes. Each epoch involves the model processing the entire dataset, adjusting its parameters, and fine-tuning its representations based on the observed patterns in the data.

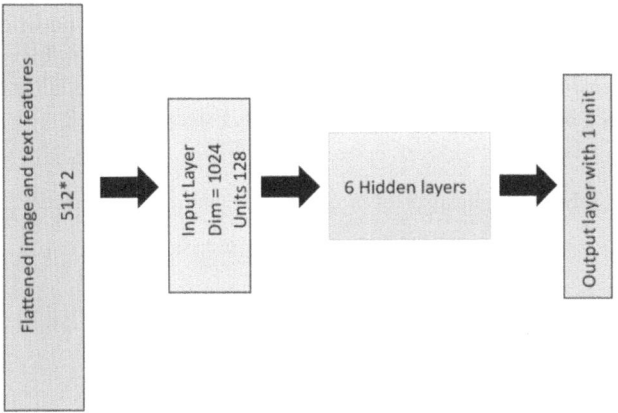

Fig. 4. Binary model architecture, fully connected layers.

The decision for the training duration is a result of a balance between achieving convergence, where the model captures the underlying patterns in the data, and mitigating the risk of overfitting, where the model becomes too tailored to the training set and loses generalization ability. Therefore, this training duration for the particular dataset encapsulates a thoughtful compromise, allowing the mCLIP model to achieve a better understanding of the dataset without overdoing the training.

4 Results

4.1 Multilevel Classification

For multilevel classification of memes, the model went testing using the SemEval 2020 Task 8 dataset. The evaluation focused on key performance metric, with a primary emphasis on the F1 macro score (average of F1 scores across all classes without considering class imbalance. A comparative analysis was conducted, juxtaposing model's results against those of several prior models documented in the SemEval 2020 Task 8 paper [1].

This comparative assessment provides insights into the effectiveness and competitiveness of model within the context of meme multilevel classification, offering a nuanced understanding of its performance in relation to other state-of-the-art models listed in the SemEval 2020 paper. The baseline model, incorporating max pooling and LSTM for text encoding and VGG-16 for image encoding, achieved an F1 score of 0.21765. In contrast, our model achieved an F1 macro score of **0.3406** outperforms the baseline model with a substantial improvement of approximately **22.5%**. A notable benchmark is provided by [2], which attains the highest F1 macro score of 0.35466. However, it's essential to analyze their approach, which utilizes two distinct strategies for classification that are unimodal and multimodal approach. In their unimodal text-only approach, they employed a feedforward neural network along with Word2Vec for text encoding, yielding an F1 macro score of 0.35466, slightly surpassing our model. In their multimodal approach, akin to ours, they deploy a multimodal bitransformer network, essentially a fusion of BERT and a ResNet-like CNN for image encoding. In this scenario, their model scores an F1 macro of 0.30. Comparing to these multimodal approaches, our model achieves a higher F1 macro score of 0.3406, listed in Table 2.

Table 2. Comparison of models having multimodal approach.

Model	Macro F1
mCLIP (ours)	0.34
MMBT	0.30
Baseline	0.21

4.2 Binary Classification

The classification of hatefulness and offensiveness involves a diverse range of models, each utilizing different datasets to demonstrate their capabilities. In the Meta Hateful Meme Challenge initiated in 2020, models were assessed and their performances, as of May 2023 according to [7] (It's noteworthy that the data utilized in this challenge were not labelled) lists the highest-scoring model as the base CLIP model, achieved a validation accuracy of 0.6492. Other models, including those fine-tuned such as VisualBERT [21], Bridge Tower [10] and some adopting unimodal approaches like ResNet [22], scored less than the base CLIP model. One standout model in this domain is Hate-CLIPper, it uses the interaction between encoded images and texts as features for training and avoided full finetuning. While these models were trained on the Hateful Meme Challenge dataset and our model is trained on SemEval 2020 dataset, it's crucial to acknowledge that direct comparisons might be challenging due to different datasets in use. However, for general understanding, mCLIP achieves an accuracy of **74.91%** in the test set binary classification, listed in Table 3. While recognizing the inherent differences in datasets our model showcases its effectiveness in binary meme classification by surpassing the benchmarks set by models trained on Meta Hateful Meme Challenge dataset.

Table 3. Comparison between CLIP and mCLIP.

Model	Accuracy percentage
CLIP	64.92
mCLIP(ours)	74.91

5 Conclusion and Future Work

In this research, we aimed to explore the complex realm of meme classification, which requires a deep understanding of both textual and visual elements. Using CLIP (Contrastive Language-Image Pre-training), we developed and trained two models, mCLIP for multilevel classification and binary classification. Our multilevel classification model, outperformed baseline models with an F1 macro score of 0.3406, showcasing competitive performance compared to models in the SemEval 2020 Task 8 paper. Additionally, our binary classification model demonstrated strong accuracy in identifying offensiveness, achieving 0.7491 accuracy.

In conclusion, our research contributes to the ongoing advancement of multimodal sentiment analysis where it shows the effectiveness of advanced models like CLIP in decoding the complexities of memes. The landscape of meme sentiment analysis is evolving, and our work represents a foundational step towards deeper exploration and refinement of models to understand the intricate interplay between text and image in memes. This research area is relatively new and requires extensive exploration before real-world applications can be realized. There is a need to develop various models that can efficiently interact with both image and text modalities and possess a deep understanding of memes. Additionally, domain-specific models trained on large datasets containing millions of memes are essential for effective meme detection and classification. Such models, coupled with well-curated task-specific datasets, will facilitate the automation of detection and enable deployment in real-world environments.

For future work, there are several avenues to explore that could further improve our model's capabilities. One avenue involves expanding the training dataset to include a larger corpus, ideally containing 50,000 to 70,000 image-text pairs. This would help refine and enhance the model's understanding by capturing a wider range of meme nuances and improving its generalization abilities. Another potential area is to specialize the model for detecting memes in specific classes, such as Islamophobia and racism. Training the model on datasets curated specifically for these content types could aid in creating a tool for targeted detection and analysis. Additionally, exploring the adaptability of our models to handle memes in multiple languages is another promising direction for future research. This would involve training the models on diverse multilingual datasets to ensure they perform well across various linguistic and cultural contexts.

References

1. Sharma, C., et al.: SemEval-2020 task 8: memotion analysis- the visuo-lingual metaphor! In: Herbelot, A., Zhu, X., Palmer, A., Schneider, N., May, J., Shutova, E. (eds.) Proceedings of the Fourteenth Workshop on Semantic Evaluation, pp. 759–773 (2020). https://doi.org/10.18653/v1/2020.semeval-1.99
2. Keswani, V., Singh, S., Agarwal, S., Modi, A.: IITK at SemEval-2020 task 8: unimodal and bimodal sentiment analysis of internet memes. In: Herbelot, A., Zhu, X., Palmer, A., Schneider, N., May, J., Shutova, E. (eds.) Proceedings of the Fourteenth Workshop on Semantic Evaluation, pp. 1135–1140 (2020). https://doi.org/10.18653/v1/2020.semeval-1.150
3. Guo, Y., Huang, J., Dong, Y., Xu, M.: Guoym at SemEval-2020 task 8: ensemble-based classification of visuo-lingual metaphor in memes. In: Proceedings of the Fourteenth Workshop on Semantic Evaluation, pp. 1120–1125 (2020)
4. Vlad, G.-A., Zaharia, G.-E., Cercel, D.-C., Chiru, C., Trausan-Matu, S.: UPB at SemEval-2020 task 8: joint textual and visual modeling in a multi-task learning architecture for memotion analysis. In: Herbelot, A., Zhu, X., Palmer, A., Schneider, N., May, J., Shutova, E. (eds.) Proceedings of the Fourteenth Workshop on Semantic Evaluation, pp. 1208–1214) (2020). https://doi.org/10.18653/v1/2020.semeval-1.160
5. Gupta, A., Kataria, H., Mishra, S., Badal, T., Mishra, V.: BennettNLP at SemEval-2020 task 8: multimodal sentiment classification using hybrid hierarchical classifier. In: Herbelot, A., Zhu, X., Palmer, A., Schneider, N., May, J., Shutova, E. (eds.) Proceedings of the Fourteenth Workshop on Semantic Evaluation, pp. 1085–1093 (2020). https://doi.org/10.18653/v1/2020.semeval-1.143
6. Morishita, T., Morio, G., Horiguchi, S., Ozaki, H., Miyoshi, T.: Hitachi at SemEval-2020 task 8: simple but effective modality ensemble for meme emotion recognition. In: Herbelot, A., Zhu, X., Palmer, A., Schneider, N., May, J., Shutova, E. (eds.) Proceedings of the Fourteenth Workshop on Semantic Evaluation, pp. 1126–1134 (2020). https://doi.org/10.18653/v1/2020.semeval-1.149
7. Zhao, B., Zhang, A., Watson, B., Kearney, G., Dale, I.: A review of vision-language models and their performance on the hateful memes challenge. arXiv preprint arXiv:2305.06159 (2023)
8. Kiela, D., et al.: The hateful memes challenge: Detecting hate speech in multimodal memes. Adv. Neural. Inf. Process. Syst. **33**, 2611–2624 (2020)
9. Radford, A., et al.: Learning transferable visual models from natural language supervision. In: International Conference on Machine Learning, pp. 8748–8763. PMLR (2021)
10. Xu, X., Wu, C., Rosenman, S., Lal, V., Che, W., Duan, N.: Bridgetower: building bridges between encoders in vision-language representation learning. In: Proceedings of the AAAI Conference on Artificial Intelligence, vol. 37, no. 9, pp. 10637–10647 (2023)
11. Tung, P.T.H., Viet, N.T., Anh, N.T., Hung, P.D.: SemiMemes: a semi-supervised learning approach for multimodal memes analysis. In: International Conference on Computational Collective Intelligence, pp. 565–577. Springer Nature Switzerland, Cham (2023). https://doi.org/10.1007/978-3-031-41456-5_43
12. Yang, Y., Wang, K.T., Zhan, D.C., Xiong, H., Jiang, Y.: Comprehensive semi-supervised multi-modal learning. In: IJCAI, pp. 4092–4098 (2019)
13. Hakimov, S., Cheema, G.S., Ewerth, R.: TIB-VA at SemEval-2022 Task 5: a multimodal architecture for the detection and classification of misogynous memes. In: Emerson, G., et al. (eds.) Proceedings of the 16th International Workshop on Semantic Evaluation (SemEval-2022), pp. 756–760 (2022). https://doi.org/10.18653/v1/2022.semeval-1.105

14. Kumar, G.K., Nandakumar, K.: Hate-CLIPper: Multimodal hateful meme classification based on cross-modal interaction of CLIP features. In: Biester, L., et al. (eds.), Proceedings of the Second Workshop on NLP for Positive Impact (NLP4PI), pp. 171–183 (2022). https://doi.org/10.18653/v1/2022.nlp4pi-1.20
15. Jin, W., Lance, W.: The hateful memes challenge next move. arXiv preprint arXiv:2212.06655 (2022)
16. Smitha, E.S., Sendhilkumar, S., Mahalaksmi, G.S.: Meme classification using textual and visual features. In: Hemanth, D., Smys, S. (eds.) Computational Vision and Bio Inspired Computing . LNCVB, vol 28. Springer, Cham (2018). https://doi.org/10.1007/978-3-319-71767-8_87
17. Aggarwal, A., et al.: Two-way feature extraction using sequential and multimodal approach for hateful meme classification. Complexity **2021**, 1–7 (2021)
18. Amalia, A., Sharif, A., Haisar, F., Gunawan, D., Nasution, B.B.: Meme opinion categorization by using optical character recognition (OCR) and naïve Bayes algorithm. In: 2018 Third International Conference on Informatics and Computing (ICIC), pp. 1–5. IEEE (2018)
19. Nayak, R., Kannantha, B.U., Gururaj, C.: Multimodal offensive meme classification using transformers and BiLSTM. Int. J. Eng. Adv. Technol. **11**(3), 96–102 (2022)
20. SemEval-2020 Task 8 dataset. Accessed on 14 Sept 2023
21. Li, L.H., et al.: VisualBERT: a simple and performant baseline for vision and language. arXiv preprint arXiv:1908.03557 (2019)
22. He, K., Zhang, X., Ren, S., Sun, J.: Deep residual learning for image recognition. In: Proceedings of the IEEE Conference on Computer Vision and Pattern Recognition, pp. 770–778 (2016)
23. Sharma, P., Ding, N., Goodman, S., Soricut, R.: Conceptual captions: a cleaned, hypernymed, image alt-text dataset for automatic image captioning. In: Proceedings of the 56th Annual Meeting of the Association for Computational Linguistics (Volume 1: Long Papers), pp. 2556–2565 (2018)

Enhancing Marine Litter Management in the Gulf of Aqaba Through AI

Mohammad Wahsha[1]( ), Heider Wahsheh[2], and Tariq Al-Najjar[3]

[1] Marine Science Station, The University of Jordan, Aqaba Branch, Aqaba, Jordan
m.wahsha@ju.edu.jo
[2] Department of Information Systems, College of Computer Science and Information Technology, King Faisal University, Al-Ahsa, Al Hofuf 31982, Saudi Arabia
hwahsheh@kfu.edu.sa
[3] Faculty of Basic and Marine Sciences, The University of Jordan, Aqaba Branch, Aqaba, Jordan
t.najjar@ju.edu.jo

Abstract. This paper investigates the critical issue of marine litter in the Gulf of Aqaba, a key marine ecosystem in the Red Sea known for its rich biodiversity and economic significance to tourism and fishing industries. It emphasizes the vulnerability of this semi-enclosed body of water to various types of marine litter, including plastic waste and discarded fishing gear, posing significant threats to marine life through ingestion and entanglement. To address the limitations of traditional monitoring methods, the study proposes the innovative use of Artificial Intelligence (AI) technologies, such as machine learning and computer vision, to enhance the detection, quantification, and categorization of marine debris. The paper aims to develop an advanced prototype for underwater image analysis using AI, designed to significantly contribute to environmental preservation efforts by providing efficient, real-time assessments of marine litter in the Gulf of Aqaba, demonstrating the potential of AI to transform marine conservation strategies.

Keywords: Artificial Intelligence · K-Nearest Neighbors · Environmental Monitoring · Marine Litter · Convolutional Neural Networks · Gulf of Aqaba

1 Introduction

Marine litter poses a significant threat to global oceanic ecosystems, including the Gulf of Aqaba, a northeastern part of the Red Sea known for its exceptional biodiversity, vibrant coral reefs, diverse marine species, and crystal clear waters [1]. This area now confronts the escalating menace of marine litter, which compromises both its ecological balance and economic potential, notably impacting tourism and fishing industries [2]. The sources and types of marine litter are diverse, ranging from plastic waste to discarded fishing gear, all contributing to the degradation of marine habitats and endangering aquatic life [3].

The Gulf of Aqaba, due to its semi-enclosed nature and unique hydrodynamic conditions, is especially vulnerable to marine litter accumulation, posing a significant hazard to marine organisms through ingestion and entanglement [4]. Research highlights the predominance of plastic waste, exacerbating the litter issue further fueled by the region's tourism appeal. Coral reefs, critical to the area's marine biodiversity, face dire threats from physical damage and potential mortality due to toxic plastics [5]. Focused studies on the Jordanian coast of the Gulf, such as those by [1], as well as [2] and [6], have explored the nature, extent, and composition of both shore and submerged marine litter, identifying metal cans, plastics, and glass as prevalent types. These studies provide essential insights for policy and management, demonstrating the human impact on marine environments.

To address the Gulf of Aqaba's marine litter complexity, we need innovative monitoring methods. Traditional approaches often fall short, failing to provide timely and extensive data collection. Artificial Intelligence (AI) technologies, including machine learning and computer vision, offer promising solutions for enhancing marine litter management [7]. They are poised to transform debris detection, quantification, and categorization through aerial and satellite imagery, underwater drones, and stationary coastal and offshore cameras [8].

AI implementation allows for the training of algorithms to accurately identify debris types across various marine environments, facilitating quick assessments of litter accumulation and distribution [9]. This method enhances monitoring efficiency, helps pinpoint pollution sources and hotspots, and supports targeted interventions and informed policymaking [8].

Moreover, AI-driven systems can be integrated into existing marine management frameworks, improving decision-making by providing precise, real-time litter data [9]. These advancements support cleanup prioritization, resource optimization, and impact assessment of mitigation efforts. AI further bolsters citizen science initiatives, encouraging community participation in conservation through apps that allow users to identify and map debris [10].

This study aims to leverage AI for developing an advanced prototype for underwater image analysis, focusing on detecting and assessing marine litter. Such technology is vital for areas like the Gulf of Aqaba, where enhanced monitoring can significantly contribute to environmental preservation efforts.

2 Materials and Methods

This study aimed to develop a prototype for monitoring the health of marine environments through underwater image analysis, employing K-Nearest Neighbor (K-NN) complemented by feature extraction techniques, notably the histogram orientation method, and Convolutional Neural Network (CNN) models. The methodology encompassed several integrated steps:

Data Collection: Between 2023 and 2024, we captured 21.6K high-resolution underwater images using a GoPro camera along the Jordanian side of the Gulf of Aqaba. This collection included areas impacted by marine debris and pristine locations. This strategy ensured the models were trained and tested across a broad spectrum of conditions, as recommended by [9] and [10], enhancing their capability to accurately classify and assess marine waste.

Data Preprocessing: Collected images underwent a series of standardization and normalization processes, such as resizing to uniform dimensions, contrast adjustment, and noise reduction application. This preprocessing was essential for reducing variability unrelated to the analysis, preparing the data for effective feature extraction and model training.

Feature Extraction Using Histogram Orientation Method: The histogram orientation method was applied to the preprocessed images to calculate the distribution of gradient orientations. This technique was critical for extracting significant features that distinguished between polluted and unpolluted marine environments, thereby improving the K-NN model's accuracy in classification.

Model Development: A K-NN model was implemented using the extracted features to classify images based on the similarity of their feature histograms to those in the training set. This approach evaluated the effectiveness of integrating traditional machine learning with advanced feature extraction methods in identifying marine waste. A CNN model was also developed, trained on the raw, preprocessed images. This model autonomously detected and learned relevant features from the image data, leveraging deep learning's capabilities for automatic feature detection and classification, offering a comparison to traditional methods.

Model Training and Validation: The models were trained on a designated dataset and validated on a separate set using cross validation methods to ensure their robustness. This phase was pivotal for optimizing model parameters for maximum performance and preventing overfitting, ensuring the models' applicability to novel, unseen data.

Performance Evaluation: The effectiveness of the models was assessed using metrics such as accuracy, precision, recall, and F1-score, among others relevant to image classification tasks. This quantitative analysis aimed to identify the most efficient model in accurately classifying marine waste within underwater images.

Prototype Development: The model demonstrating superior performance was incorporated into a prototype system designed for real-time marine waste assessment. This phase translated the research findings into a practical application, facilitating ongoing monitoring and health assessment of marine environments.

Ethical Considerations and Environmental Impac: The research adhered to ethical guidelines for environmental research, aiming to minimize the ecological footprint during data collection ensuring minimal adverse effects on marine ecosystems in compliance with conservation principles.

3 Results

The evaluation begins with insights derived from the performance metrics detailed in Table 1, revealing the CNN model's precision in discerning between marine waste and healthy marine environments. The True Positive Rate (TPR), or Recall, highlights the model's proficiency, with an 80% success rate in identifying marine waste and an even more impressive 90% accuracy for healthy marine instances. These figures not only demonstrate the model's effectiveness in distinguishing between the two classes but also underscore its potential in guiding conservation strategies with precision. A deeper dive into the False Positive Rate (FPR) and precision metrics, as presented in the same table, further cements the model's utility. The FPR values of 0.100 for marine waste and 0.200 for healthy marine illustrate the model's stringent filter against false alarms, pivotal in minimizing unwarranted interventions. Precision scores of 0.889 for marine waste and 0.818 for healthy marine reflect the model's high accuracy rate in positive predictions, bolstering its reliability. The F-Measure and Matthews Correlation Coefficient (MCC), also detailed in the initial table, validate the model's balanced performance. The F-Measure, offering a harmonic mean of precision and recall, along with the MCC, which evaluates the model's quality across true and false positives and negatives, underscore the model's robustness. These metrics, with F-Measures of 0.800 for marine waste and 0.900 for healthy marine, alongside MCC scores of 0.842 and 0.857 respectively, indicate a strong positive correlation between observed and predicted classifications, highlighting the model's efficacy.

Table 1. Performance Metrics of CNN in Marine Waste Identification

Class	TPR (Recall)	FPR	Precision	F-Measure	MCC
Marine waste	0.800	0.100	0.889	0.800	0.842
Healthy	0.900	0.200	0.818	0.900	0.857
Weighted Average	0.850	0.150	0.854	0.850	0.850

Transitioning to the model's classification accuracy and error analysis, presented in Table 2, an 85% rate of correctly classified instances underscores the model's overarching accuracy. This high accuracy rate, coupled with a mere 15% of instances incorrectly classified, affirms the model's effectiveness. The Kappa Statistic, at 0.7, further indicates substantial agreement between the model's predictions and the actual classifications, reinforcing its consistency and reliability. Error metrics, including the Mean Absolute Error (MAE) and Root Mean Squared Error (RMSE), with values of 0.1895 and 0.3828 respectively, highlight the model's minimal prediction errors, a testament to its precision. The Relative Absolute Error (RAE) and Root Relative Squared Error (RRSE), though indicating areas for improvement, still emphasize the model's significant predictive power over simpler baseline models, as elucidated in Table 2. The meticulous examination of the CNN model's performance metrics and classification accuracy, as articulated through the data in both Tables 1 and 2, elucidates its dominance and efficacy in marine waste assessment. The high accuracy, balanced precision and recall, alongside minimal error rates, position the CNN model as a robust and reliable tool in the environmental monitoring landscape, demonstrating its indispensable value in addressing marine conservation challenges.

Table 2. CNN Model's Accuracy and Predictive Error Insights

Metric	Value
Correctly Classified Instances	85%
Incorrectly Classified Instances	15%
Kappa Statistic	0.7
MAE	0.1895
RMSE	0.3828
RAE	37.8929%
RRSE	76.5545%

Within the intricate domain of marine waste assessment, the Convolutional Neural Network (CNN) model emerges as a paramount tool, showcasing its adeptness through a detailed analysis of performance metrics and classification accuracy. This comprehensive review leverages data from two critical tables to underscore the model's capabilities and its consequential role in bolstering environmental monitoring initiatives.

The K-NN model's performance metrics, as outlined in Table 3, reveal a TPR of 0.800 for marine waste, mirroring that of the CNN model. However, the TPR for healthy marine instances drops to 0.600, indicating a less robust ability of the K-NN model to accurately classify healthy marine environments compared to the CNN model's 0.900 TPR. Furthermore, the FPR for the K-NN model stands at 0.400 for marine waste and 0.200 for healthy marine, showcasing a higher likelihood of false alarms compared to the CNN model's lower FPR values.

Precision, F-Measure, and MCC further differentiate the two models' performance. The K-NN model achieves a precision of 0.667 for marine waste and 0.750 for healthy marine, which, while respectable, falls short of the CNN model's precision scores. Similarly, the F-Measure and MCC values for the K-NN model, although solid, do not reach the high benchmarks set by the CNN model, underscoring a comparative deficit in efficacy and reliability in marine waste assessment.

Table 3. K-NN Model Efficacy in Marine Environmental Classification

Class	TPR (Recall)	FPR	Precision	F-Measure	MCC
Marine waste	0.800	0.400	0.667	0.800	0.727
Healthy	0.600	0.200	0.750	0.600	0.667
Weighted Average	0.700	0.300	0.708	0.700	0.697

The classification accuracy and error analysis provided in Table 4 for the K-NN model show a correctly classified instance rate of 70%, significantly lower than the CNN model's 85% accuracy rate. This discrepancy highlights the CNN model's superior capability in accurately identifying marine waste and healthy marine conditions.

Additionally, the K-NN model's 30% rate of incorrectly classified instances, coupled with a lower Kappa Statistic of 0.4, indicates a moderate level of agreement and a higher error propensity compared to the CNN model's outcomes.

Table 4. Accuracy and Error Evaluation for the K-NN Model

Metric	Value
Correctly Classified Instances	70%
Incorrectly Classified Instances	30%
Kappa Statistic	0.4
MAE	0.3554
RMSE	0.4511
RAE	71.0714%
RRSE	90.2236%

Error metrics such as MAE, RMSE, and their relative counterparts further illustrate the differences in prediction accuracy between the two models. The K-NN model exhibits higher values in both MAE and RMSE compared to the CNN model, indicating larger average errors in its predictions. The RAE and RRSE for the K-NN model also surpass those of the CNN model, suggesting a greater magnitude of error relative to a simple baseline model.

The comprehensive comparison of performance metrics and classification accuracy detailed in Table 5 unequivocally demonstrates the CNN model's dominance over the K-NN model in the realm of marine waste assessment. This analysis reveals the CNN model's superior precision, higher accuracy, and lower error rates, positioning it as a notably more robust and reliable tool for environmental monitoring. The CNN model excels in accurately identifying marine waste and healthy marine environments, as evidenced by its superior true positive rates and notably lower false positive rates for marine waste, alongside its more favorable error metrics. This prowess is further exemplified by its balanced precision and recall, as highlighted by the F-Measure, and a strong positive correlation between observed and predicted classifications, as denoted by the MCC.

These findings not only affirm the CNN model's enhanced capability in classifying marine waste and healthy marine instances but also its preferential utility in applications demanding high precision and reliability. Such comprehensive performance superiority underscores the CNN model's indispensability in marine conservation efforts, clearly establishing it as the preferable choice for achieving high precision and minimal error in environmental monitoring and conservation initiatives.

Table 5. Comparative Analysis Table: CNN vs. K-NN Model Performance in Marine Waste Assessment

Metric	Category	CNN Model	K-NN Model	Remarks
Classification Accuracy	-	85%	70%	CNN exhibits superior accuracy in classification
TPR	Marine Waste	80%	80%	Both models perform equally for marine waste
	Healthy Marine	90%	60%	CNN is more effective in identifying healthy marine environments
FPR	Marine Waste	10%	40%	CNN shows a lower likelihood of false alarms for marine waste
	Healthy Marine	20%	20%	Both models have the same FPR for healthy marine
Precision	Marine Waste	88.9%	66.7%	CNN demonstrates higher precision for marine waste
	Healthy Marine	81.8%	75%	CNN also shows higher precision for healthy marine
F-Measure	Marine Waste	80%	80%	Equal F-Measure for marine waste, indicating similar balance of precision and recall
	Healthy Marine	90%	60%	CNN has a better harmonic mean of precision and recall for healthy marine
MCC	Marine Waste	0.842	0.727	CNN indicates a stronger positive correlation for marine waste
	Healthy Marine	0.857	0.667	CNN shows better correlation for healthy marine as well

(*continued*)

Table 5. (*continued*)

Metric	Category	CNN Model	K-NN Model	Remarks
Kappa Statistic	-	0.7	0.4	CNN exhibits substantial agreement between predictions and actual classifications
MAE	-	0.1895	0.3554	CNN has lower average prediction errors
RMSE	-	0.3828	0.4511	CNN demonstrates lower errors in predictions
RAE	-	37.8929%	71.0714%	CNN shows more accurate predictions relative to a baseline
RRSE	-	76.5545%	90.2236%	CNN exhibits lower relative squared errors, indicating better performance

4 Discussion

In the Gulf of Aqaba, a substantial portion of marine litter, predominantly composed of plastic materials, originates from recreational and maritime activities. This includes significant contributions from ferry operations and activities at the port near MSS beach, with an estimated tens of millions of items entering the water each year. Shipping and port activities are major contributors to this issue, whereas the fishing industry adds a relatively small fraction to the overall marine debris. Such pollution severely threatens the health of marine ecosystems [11]. In response, ASEZA has implemented a zero discharge policy, mandating comprehensive environmental impact assessments for all significant development projects to mitigate potential adverse effects on the marine environment. This approach includes penalties for any environmental harm caused by these activities, underlining a strong dedication to environmental preservation [11–13]. Despite ongoing efforts and annual clean-up initiatives, the continuous rise in marine litter highlights the critical need for sustained and intensified measures to safeguard this essential and vibrant marine ecosystem in the Red Sea from further degradation.

The region of Aqaba stands out for its remarkable biodiversity and the thriving coral reefs that adorn its waters, presenting a natural wonder that attracts admiration and study [14]. However, this precious marine environment is under threat from increasing pollution levels, particularly from an assortment of debris, including plastics and abandoned fishing gear [15]. Such pollutants pose severe risks to the ecological balance and economic health of the area, especially affecting critical sectors such as tourism and fisheries [16, 17]. These impacts are substantiated by numerous studies [2, 6], highlighting the pressing need for concerted action. Additionally, the Gulf of Aqaba's distinct geographical and hydrodynamic characteristics make it particularly prone to litter accumulation. This condition significantly heightens the danger to marine life, exposing numerous species to the hazards of ingestion and entanglement [5]. The situation demands a reinforced commitment to environmental preservation and proactive measures to combat pollution, ensuring the protection and sustainability of this unique marine ecosystem.

Confronting the intricate issue of marine litter in the Gulf of Aqaba calls for innovative and advanced monitoring techniques beyond traditional methodologies, which fall short in providing timely and exhaustive data collection. This study ventures into the realm of Artificial Intelligence (AI), harnessing machine learning and computer vision to revolutionize debris detection, classification, and quantification through cutting-edge tools such as aerial and satellite imagery, underwater drones, and fixed coastal and offshore cameras [7, 10]. AI's capability to train algorithms for accurate debris identification across various marine settings significantly enhances monitoring precision, facilitating the identification of pollution sources and accumulation hotspots, thereby enabling targeted remedial actions and informed policy development [17, 18]

This research corroborates previous findings [19–21], affirming the superiority of the Convolutional Neural Network (CNN) model over the K-Nearest Neighbor (K-NN) model in the precise classification of marine waste and the delineation of healthy marine environments. The CNN model, through its deep learning architecture, excels in autonomously learning and identifying pertinent features from raw image data, thus efficiently capturing complex patterns in marine litter. This model's hierarchical feature extraction capability and adeptness in handling the nonlinear intricacies of underwater imagery ensure heightened accuracy and reliability in classification tasks [22].

This study addresses several limitations inherent in existing monitoring and management practices of marine litter, particularly in the Gulf of Aqaba's context:

- Traditional Monitoring Methods' Shortcomings: Previous approaches often proved insufficient for the timely and comprehensive data collection essential for effective marine litter management. The vastness and complexity of marine settings like the Gulf of Aqaba rendered conventional methods inadequate. The proposed AI-based approach, leveraging machine learning and computer vision, surpasses these limitations, offering advanced capabilities for debris detection, quantification, and classification.
- Efficiency in Identifying Debris Types: The ability to accurately identify various marine debris types across different environments is crucial for effective monitoring and management. The AI technologies employed in this study facilitate the training of algorithms capable of precise debris identification in diverse marine settings, thus enhancing monitoring efficiency and supporting the development of targeted interventions and informed policy decisions.
- Integration with Marine Management Frameworks: Existing methods lacked sufficient integration with marine management frameworks to enhance decision-making through accurate, real-time litter data. The AI-driven systems proposed here can be seamlessly incorporated into existing frameworks, significantly enhancing decision-making capabilities by providing precise, real-time marine litter data, thereby aiding

in cleanup prioritization, resource allocation, and the evaluation of mitigation efforts' impacts.
- Accuracy and Efficiency in Marine Waste Assessment: Employing a CNN model for underwater image analysis markedly improves performance over traditional methods like the K-NN model in accurately classifying marine waste and pristine marine environments. The CNN model's ability to autonomously detect and learn relevant features from image data boosts classification accuracy and operational efficiency, thus significantly enhancing marine litter monitoring and management, especially in ecologically sensitive and economically significant regions like the Gulf of Aqaba.

The significance of these advancements extends far beyond academic inquiry, offering practical solutions for marine litter management and conservation within the Gulf of Aqaba. The demonstration of AI technology, particularly CNN-based methods, in the assessment of marine litter emphasizes the critical need for integrating advanced technological solutions into environmental monitoring frameworks. Such integration enables policymakers and managers to refine decision-making processes, prioritize cleanup efforts more effectively, and allocate resources more efficiently, resonant with the insights and recommendations presented by.[22–24].

AI's capabilities for real-time analysis herald a new era in marine litter monitoring and response strategies. Continuous evaluation of underwater imagery by AI systems provides immediate insights into litter distribution and accumulation, enabling proactive approaches like targeted cleanups and pollution source tracing. This proactive stance not only mitigates ecological damage but also lessens the economic impacts on crucial industries such as tourism and fishing, marking a significant stride towards sustainable marine ecosystem management.

5 Conclusions

The study underscores the critical need for innovative monitoring and assessment tools in combating marine debris, with AI technologies offering promising solutions. The superior performance of the CNN model in marine waste assessment exemplifies the potential of AI in enhancing environmental monitoring efforts, paving the way for more effective conservation strategies in the Gulf of Aqaba and beyond. The integration of AI-driven systems into marine management frameworks represents a significant step forward in safeguarding marine ecosystems, emphasizing the importance of technological innovation in environmental conservation. Future research should explore integrating multi-modal data, including acoustic and environmental sensors, to enhance AI models' predictive power for more holistic marine ecosystem assessments. Expanding data sources to include satellite imagery and unmanned aerial vehicles could improve the scale and detail of monitoring efforts. Additionally, developing AI-based decision support systems tailored to local stakeholders' needs promises valuable advancements. Collaborating with end-users, such as coastal communities and conservation organizations, ensures that technological innovations are aligned with local knowledge, cultural values, and management priorities, fostering sustainable marine conservation practices.

References

1. Abu-Hilal, A., Al-Najjar, T.: Litter pollution on the Jordanian shores of the Gulf of Aqaba (Red Sea). Mar. Environ. Res. **58**(1), 39–63 (2004)
2. Al-Najjar, T., Al-Shiyab, H.: Marine litter at (Al-Ghandoor area) the most northern part of the Jordanian coast of the Gulf of Aqaba. Red. Sea. Natural Sci. **3**, 921–926 (2011)
3. Abdel-Halim, A., et al.: Environmental studies on the Aqaba Gulf Coastal Waters during 2011–2013. J. Environ. Prot. **7**, 1411–1437 (2016)
4. Wahsha M AS, Juhmani A, Buosi A, Sfriso A, Sfriso A.: Assess the environmental health status of macrophyte ecosystems using an oxidative stress biomarker. Case studies: The Gulf of Aqaba and the Lagoon of Venice. Energy Procedia **125**, 19–26 (2017)
5. Vered, G., Shenkar, N.: Plastic pollution in a coral reef climate refuge: occurrence of anthropogenic debris, microplastics, and plasticizers in the Gulf of Aqaba. Sci. Total. Environ. **905**, 167791 (2023)
6. Abu-Hilal, A., Al-Najjar, T.: Marine litter in coral reef areas along the Jordan Gulf of Aqaba, Red Sea. J. Environ. Manage. **90**(2), 1043–1049 (2009)
7. Anthony, D., et al.: Trends in marine pollution mitigation technologies: scientometric analysis of published literature (1990–2022). Reg. Stud. Mar. Sci. **66**, 103156 (2023)
8. Politikos, D.V., Adamopoulou, A., Petasis, G., Galgani, F.: Using artificial intelligence to support marine macrolitter research: a content analysis and an online database. Ocean Coast. Manag. **233**, 106466 (2023)
9. Seyyedi, S.R., Kowsari, E., Ramakrishna, S., Gheibi, M., Chinnappan, A.: Marine plastics, circular economy, and artificial intelligence: a comprehensive review of challenges, solutions, and policies. J. Environ. Manage. **345**, 118591 (2023)
10. Veerasingam, S., Chatting, M., Asim, F.S., Al-Khayat, J., Vethamony, P.: Detection and assessment of marine litter in an uninhabited island, Arabian Gulf: a case study with conventional and machine learning approaches. Sci. Total. Environ. **838**(2), 156064 (2022)
11. UNDP. (2023). Aqaba Marine Reserve Management Plan 2022 - 2026. Homepage, https://www.undp.org. Accessed 15 Feb 2024
12. UNDP. (2014). Jordan ICZM Country Report 2014. Homepage, https://www.undp.org/jordan/publications. Accessed 14 Feb 2024
13. UNDP. (2015). State of the Coast Environment, Report for Aqaba, 2015. Homepage, https://jo.chm-cbd.net. Accessed 16 Feb 2024
14. Kteifan, M., Wahsha, M., Al-Horani, F.A.: Assessing stress response of Stylophora pistillata towards oil and phosphate pollution in the Gulf of Aqaba, using molecular and biochemical markers. Chem. Ecol. **33**(4), 281–294 (2017)
15. Khalaf, M.A., Ma'ayta, S.S., Wahsha, M., Manasrah, R.S., Al-Najjar, T.H.: Community structure of the deep-sea fishes in the northern Gulf of Aqaba, Red Sea (Osteichthyes and Chondrichthyes). Zool. Middle East **65**(1), 40–50. (2019)
16. Al-Absi, E., Manasrah, R., Abukashabeh, A., Wahsha, M.: Assessment of heavy metal pollutants at various sites along the Jordanian coastline of the Gulf of Aqaba, Red Sea. Int. J. Environ. Anal. Chem. **99**(8), 726–740 (2019)
17. Jia, T., et al.: Deep learning for detecting macroplastic litter in water bodies: a review. Water Res. **231**, 119632 (2023)
18. Isabelle, D.A., Westerlund, M.: A review and categorization of artificial intelligence-based opportunities in wildlife. Ocean Land Conserv. Sustain. **14**, 1979 (2022)
19. Arnita, A., Yani, M., Marpaung, F., Hidayat, M.,'Widianto, A.: A comparative study of convolutional neural network and k-nearest neighbours algorithms for food image recognition. Comput. Technol. **27**(6), 88–99 (2022)

20. Altini, N., et al.: Segmentation and identification of vertebrae in CT scans using CNN, k-means clustering and k-NN. Informatics **8**, 40 (2021)
21. Mucherino, A., Papajorgji, P.J., Pardalos, P.M.: Data Mining in Agriculture. Springer Optimization and Its Applications, vol 34. Springer, New York, NY (2009). https://doi.org/10.1007/978-0-387-88615-2
22. Wahsheh, H.A.M., Wahsha, M.: From farm to algorithms: AI-infused aquaculture & natural antidotes as a game-changer in disease mitigation. In: Proceedings of the 14th International Conference on Information and Communication Systems (ICICS), pp. 1–6. (2023)
23. Wahsha, M., Wahsheh, H.A.M., Hayek, W., Al-Tarawneh, H., Khalaf, M., Al-Najjar, T.: Bioinformatics research through image processing of histopathological response to stonefish venom. Int. J. Adv. Comput. Sci. Appl. **12**(11), 258–263 (2021)
24. Wahsheh, H., Wahsha, M., Al-Najjar, T., Khalaf, M., Al-Tarawneh, H.: Analyzing hepatotoxicity of marine venoms using artificial intelligence: an ecoinformatics perspective. In: Proceedings of the 24th International Arab Conference on Information Technology (ACIT), pp. 1–7. Ajman, United Arab Emirates (2023)

Solving the Rush Hour Puzzle Problem by Different Heuristics

Yannick Schmid[1], Rolf Dornberger[2], and Thomas Hanne[1(✉)]

[1] University of Applied Sciences and Arts Northwestern Switzerland, Olten, Switzerland
thomas.hanne@fhnw.ch
[2] Institute for Information Systems, University of Applied Sciences and Arts Northwestern Switzerland, Basel, Switzerland

Abstract. This research focuses on modeling and optimization of the Rush Hour puzzle, a grid-based board game whose objective is to determine a shortest sequence of movements of cars to let the red car exit a crowded parking lot. Recognized for its PSPACE-complete complexity, the Rush Hour problem presents significant challenges. The study explores Breadth-First Search (BFS) and A* search algorithms with various heuristics within the Subgoal framework. The implementations are evaluated against the 10,000 most complex Rush Hour configurations. Results demonstrate that the A* search algorithm markedly decreases the count of nodes explored during the solution-finding process. The heuristics developed in this study, along with the rush hour solver, provides possibilities for refining the subgoal search algorithm in a later step.

Keywords: Rush Hour problem · Rush Hour puzzle · heuristic · Subgoal Search · A* · BFS

1 Introduction

Rush Hour, a physical puzzle designed by Nob Yoshigahara in 1970, is played on a 6 × 6 board with several cars (2 × 1 units in size) and trucks (3 × 1 units in size). Vehicles can move either forward or backward, but not sideways. The board features a single exit door, shown in Fig. 1. The objective is to iteratively maneuver the vehicles back and forth to create a feasible shortest iteration sequence of movements of all cars for the red car (2 × 1) to exit.

The generalized Rush Hour problem, which allows for an arbitrary m × n grid size and the placement of the exit at any point along the grid's perimeter, has been proven to be PSPACE-complete and therefore makes it interesting to use it as benchmark for various solver algorithms [1,6].

Fig. 1. Example of a Rush Hour puzzle [8].

2 Problem Description and Solution Methods

2.1 Problem Description

1) **Board Representation:**
The Rush Hour puzzle is played on an m × n grid, with a standard commercial version using a 6x6 grid. The grid represents a parking lot filled with cars (size 2x1) and trucks (size 3x1). Each vehicle is placed either horizontally or vertically within the grid. The grid also has a designated exit on its perimeter.

2) **Vehicles:**
Each vehicle is characterized by its size, orientation, and position on the grid. Cars have a size of 2x1 units, while trucks have a size of 3x1 units. A vehicle's orientation can be either horizontal or vertical, and its position is determined by the coordinates of its top-left cell on the grid.

3) **Objective:**
The goal of the Rush Hour puzzle is to move the vehicles in the shortest sequence that creates a clear path for the red car to exit the parking lot through the designated exit.

4) **Moves:**
A move consists of shifting a single vehicle either forward or backward in the direction of its orientation (i.e., horizontal vehicles can move left or right, and vertical vehicles can move up or down). Vehicles are restricted to their respective rows or columns and cannot change their orientation. Vehicles cannot move outside the grid boundaries, overlap other vehicles, or share a tile with another vehicle.

5) **Constraints:**
Legal moves are constrained by the vehicles' orientations and the grid's boundaries. Each move must maintain the board's validity by ensuring no two vehicles share the same tile and that vehicles stay within the grid boundaries.
6) **Solution:**
A solution to the Rush Hour puzzle is a sequence of legal moves that enables the red car to reach the designated exit. The aim is to find the shortest possible sequence, thus the sequence with the fewest number of moves, minimizing the total moves required for the red car to exit the parking lot.

2.2 Optimization Methods

Several optimization methods have been evaluated to solve the Rush Hour puzzle in the past. Previous approaches include:

1) **Uninformed Search Algorithms:**
Breadth-First Search (BFS) and Depth-First Search (DFS) are examples of uninformed search algorithms, characterized by their exploration strategy in traversing tree or graph structures. BFS operates by exploring all nodes at the current depth level prior to progressing to the next, hence emphasizing breadth over depth. Conversely, DFS maximizes depth, exhaustively pursuing a path before backtracking to explore sibling branches [9].
2) **Heuristic Search Algorithms:**
Heuristic search algorithms like A* and IDA* (Iterative Deepening A*) incorporate heuristics to guide the search process toward a solution more efficiently. These algorithms estimate the cost of reaching the goal from the current state and prioritize states with lower estimated costs. Common heuristics for Rush Hour include the number of vehicles blocking the red car's path and the minimum number of moves required for each blocking vehicle to clear the way. There is however no perfect heuristics available today [5].
3) **Constraint Programming:**
Constraint programming is a technique that involves defining the problem in terms of variables and constraints, and then using a constraint solver (e.g., clingo or DLV) to find a solution that satisfies all constraints [2].
4) **Genetic Algorithms:**
Genetic algorithms are optimization techniques inspired by the process of natural selection. They work with a population of candidate solutions, applying genetic operations such as mutation, crossover, and selection to evolve better solutions over time. In the case of Rush Hour, the population consists of sequences of moves, and the genetic algorithm attempts to find a sequence that successfully leads the red car to the exit [5].

3 Goals and Methodology

3.1 Goals

The research builds upon the work of [3], where a Subgoal Search algorithm is used to solve complex problems suspecting high potential of this Subgoal Search algorithm for addressing the Rush Hour puzzle. Hence, this paper aims to adapt the software prototype developed by K. Czechowski et al. [3] to include a solver setup for the Rush Hour Puzzle. However, due to missing base line policies and heuristics that would allow a direct transfer of the solution, the paper enhances the basic understanding, which includes implementing an effective heuristic for training the transformer networks used to identify subgoals. Additionally, the paper provides a Rush Hour solver that operates on a simple BFS solver. This solver serves as baseline for a subsequent inclusion of the BF-KSubs implementation as described [3] which may lead to significant further improvements for complex search spaces.

The objectives of this research are as follows:

1. Implement a BFS-based solver for Rush Hour in the software environment developed by K. Czechowski et al. [3].
2. Develop an A*-based solver that supports multiple heuristics.
3. Validate both the BFS implementation and the A* heuristics by benchmarking them against the 10,000 most challenging Rush Hour problems.

3.2 Methodology

Breadth-First Search (BFS) and the A* algorithm with three distinct heuristics are compared. These algorithms are implemented in Python and used to solve the 10,000 most difficult Rush Hour initial states.

1) Software and Hardware Setup: For this project, the solution from https://github.com/subgoal-search/subgoal-search is used as a base. Necessary updates are made to ensure compatibility with Python 3.7, due to the deprecation of several dependency packages. The computational environment for running this setup is a Google Cloud VM, equipped with 16GB of RAM and 4 CPU cores.

No GPU support is utilized for the benchmarking of the BFS and A* algorithm implementations, though the setup is fully capable of accommodating this.

To facilitate the execution of the BFS and A* implementations, separate run configurations were established. Each configuration features its own config injection via GIN and is prepared to run efficiently in a high-performance computing environment.

2) Data Acquisition and Preprocessing: The initial states for the experiments are sourced from Michael Fogleman's database of hard Rush Hour configurations [4]. The 10,000 most challenging states devoid of obstacles are selected, ensuring consistent complexity across the dataset.

To conform with the way the solvers were built, these states had to be transformed first. Upper case letters denote horizontal cars and lower case letters represent vertical ones. Following the transformation, the states are inputted into the solvers. This approach produces a robust, uniform dataset, ensuring rigorous testing of solution methodologies. In Fig. 2, a sample character board is displayed, demonstrating the input used for both the BFS and A* implementations.

3) Breadth-First Search (BFS): The BFS implementation is adapted from a solution proposed by K.L. de Vries on his blog [10]. BFS is implemented as a standard graph search algorithm, enhanced with hash memorization and path storage. The implemented algorithm is initialized with a queue, starting from the initial state of the problem, and

```
[ '.',  '.',  '.',  'b',  'C',  'C' ],
[ '.',  '.',  '.',  'b',  '.',  '.' ],
[ '.',  'A',  'A',  'b',  '.',  '.' ],
[ 'e',  'D',  'D',  'D',  '.',  '.' ],
[ 'e',  '.',  '.',  '.',  '.',  '.' ],
[ 'F',  'F',  'G',  'G',  '.',  '.' ],
```

Fig. 2. Board with vehicle orientations encoded in the case dimension [10].

proceeds by dequeueing the next state. The algorithm then checks if the current state matches the goal state. If it does, it signifies the successful completion of the search. If not, it generates all possible succeeding states (see Fig. 3) and adds them to the end of the queue. This process continues until either a solution is found, or all states have been explored.

Hash memorization is employed to avoid the re-exploration of states, enhancing the efficiency of the BFS. A set of 'seen states' is maintained, and any new state generated is checked against this set. This avoids redundant computation and unnecessary expansions of the search space.

Finally, path storage functionality has been incorporated into the BFS. Instead of just storing states in the queue, paths leading to those states are stored. This allows the algorithm not just to establish the existence of a solution, but to provide the sequence of states that leads from the start to the goal state, thereby delivering a complete solution. This implementation adapts the conventional BFS, augmenting it with techniques designed to enhance efficiency and solution completeness.

```python
def bfs_search(start_state, goal_state):
    queue = [[start_state]]
    seen_states = set()
    while queue:
        path = queue.pop(0)
        if path[-1] == goal_state:
            return True
        for next_state in get_next_states(path[-1]):
            if next_state not in seen_states:
                seen_states.add(next_state)
                queue.append(path + [next_state])
    return False
```

Fig. 3. BFS proposed by K.L. de Vries [4].

In the following code snipped it is displayed how the next possible board states are determined. The method employed by [10] is following these steps as shown in Fig. 4:

1. An iteration over all squares on the board is done to identify the next letter, such as 'B'.
2. Adjacent squares were searched for the same character, to locate all instances of 'B'.
3. Depending on the vehicle's orientation, it was deter- mined whether it could potentially slide forward or backwards (or left or right).

```python
def get_next_states(board):
    processed_chars_set = set([EMPTY_SPACE])
    next_states = []
    for r in range(N):
        for c in range(N):
            char = board[r][c]
            if char not in processed_chars_set:
                processed_chars_set.add(char)
                delta_r = 0
                delta_c = 0
                is_vertical = not char.isupper()
                if is_vertical:
                    delta_r = 1
                else:
                    delta_c = 1

                # Find the extrema
                min_r, max_r = r, r
                min_c, max_c = c, c
                while min_r - delta_r >= 0 and  
                    min_c - delta_c >= 0 and  
                    board[min_r - delta_r][min_c - delta_c] == char:
                    min_r -= delta_r
                    min_c -= delta_c

                while max_r + delta_r < N and  
                    max_c + delta_c < N and  
                    board[max_r + delta_r]  
                        [max_c + delta_c] == char:
                    max_r += delta_r
                    max_c += delta_c


                if min_r - delta_r >= 0 and  
                    min_c - delta_c >= 0 and  
                    board[min_r - delta_r]  
                        [min_c - delta_c] == EMPTY_SPACE:
                    next_state = copy_board(board)
                    next_state[min_r - delta_r]  
                            [min_c - delta_c] = char
                    next_state[max_r][max_c] = EMPTY_SPACE
                    next_states.append(next_state)

                if max_r + delta_r < N and  
                    max_c + delta_c < N and  
                    board[max_r + delta_r]  
                        [max_c + delta_c] == EMPTY_SPACE:
                    next_state = copy_board(board)
                    next_state[min_r][min_c] = EMPTY_SPACE
                    next_state[max_r + delta_r]  
                            [max_c + delta_c] = char
                    next_states.append(next_state)
```

Fig. 4. Next states as proposed by K.L. de Vries [10].

4. The possible board configurations were calculated, considering the movement of the vehicle one step in either direction, and these were added to the list of possible next states.
5. Once the vehicle'B' was processed, it was recorded in a set, and all subsequent recurrences of 'B' during this pass of the board were skipped.

4) A* Algorithm: The A* search algorithm was implemented with three distinct heuristics to guide the search to- wards the optimal solution. A* is a best-first search algorithm that uses a heuristic to estimate the cost of reaching the goal from a given state, allowing it to prioritize states that are believed to be closer to the solution. The implemented version of A* is utilizing a priority queue as suggested in [7] and contains adaptions to save the solution paths. To explore the next possible states the same method as described in Fig. 5 was utilized.

```
def a_star_solver(start_state):
  def is_goal(state):
    return num_blocking_cars(state) == 0

  # priority queue
  queue = [(num_blocking_cars(start_state)
            , 0, start_state)]
  visited = set()
  parents = {}
  visited_count = 0

  while queue:
    (priority, depth, current)
      = heapq.heappop(queue)

    if board_str(current) in visited:
      continue

    if is_goal(current):
      # backtrack the solution path
      path, moves = backtrack_path(current, parents)
      return path, visited_count, depth

    visited.add(board_str(current))
    visited_count += 1

    for next_state, move
        in get_next_states(current):
      if board_str(next_state) not in visited:
        # Record the parent state and move
        parents[board_str(next_state)] =
            (current, move)
        estimated_cost_to_goal =
            num_blocking_cars(next_state) +depth
        heapq.heappush(queue,
            (estimated_cost_to_goal,
            depth + 1, next_state))
                        return None, visited_count, None
```

Fig. 5. A* implementation.

In the following the heuristics are explained in detail:

a) H0: Zero Heuristic: The Zero Heuristic, or H0, always returns a value of 1. It serves two purposes: first, it acts as a benchmark for comparing other heuristics; second, it checks our implementation against the results from a Breadth-First Search (BFS) algorithm.

```
def num_blocking_cars_advanced(board):
    # Find the row and column of the goal car
    for i in range(N):
        for j in range(N):
            if board[i][j] == GOAL_CAR:
                goal_row = i
                goal_col = j
                break

    # Count the number of cars blocking
    # the goal car from exiting (Type 1)
    type_1_blocking = 0
    type_1_blocking_cars = []
    for j in range(goal_col+2, N):
        if board[goal_row][j] != EMPTY_SPACE:
            type_1_blocking += 1
            type_1_blocking_cars.
                append(board[goal_row][j])

    # Count the number of cars blocking
    # the first type cars from moving (Type 2)
    type_2_blocking = []
    for type_1_car in type_1_blocking_cars:
      for i in range(N):
      for j in range(N):
       if board[i][j] == type_1_car:
         # Check the cars above and below
         # if the blocking car is vertical
         for k in range(N):
           # Check the cars above if
           # not on the first row
           if i > 0 and
           board[k][j] != EMPTY_SPACE and
           board[k][j] != GOAL_CAR and
           board[k][j] != type_1_car and
           board[k][j] not in type_2_blocking:
               type_2_blocking.append(board[k][j])

           # Check the cars below if
           # not on the last row
           if i < N - 1 and
           board[k][j] != EMPTY_SPACE and
           board[k][j] != GOAL_CAR and
           board[k][j] != type_1_car and
           board[k][j] not in type_2_blocking:
               type_2_blocking.append(board[k][j])

    return type_1_blocking
           + len(type_2_blocking)
```

Fig. 6. Heuristic H2.

b) H1: Direct Blocking Heuristic: The Direct Blocking Heuristic, or H1, counts the cars directly blocking the goal car's path out of the maze. Since these cars must move to let the goal car out, this heuristic provides a lower bound of movements needed for an optimal solution to the problem. Thus, H1 gives a reliable estimate of the minimum number of moves needed to solve the problem.
c) H2: Multiple Blocking Heuristic: The Multiple Blocking Heuristic, or H3, goes a step beyond H1 by also considering cars that block those directly in the goal car's path. This heuristic gives a fuller picture of the problem's complexity by considering secondary blockages. This heuristic is also admissible as it never overestimates the cost to the solution. See Fig. 6 for details.

4 Results and Discussion

The analysis of the implemented search algorithms provided significant insights (see Fig. 7 and 8). The BFS algorithm demonstrated the quickest average solution time, suggesting an efficient search process. However, BFS also generated the largest average number of nodes representing solutions considered during search, indicating a substantial demand on memory resources. For solving the 6x6 board this is no issue, but thinking about the generalized rush hour problem this could quickly become an issue. In general, we, however, see a positive correlation between the number of nodes being generated and the solution time (see Fig. 9).

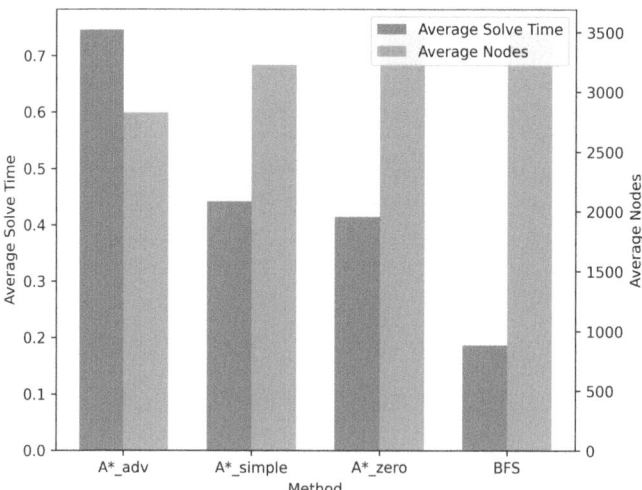

Fig. 7. Average number of nodes and average solution time the four considered methods, A* adv, A* simple, A* zero, and BFS.

The two A* variants, A* simple and A* adv, showed a balanced performance in terms of average solution time and the number of nodes generated. Although these algorithms had slower average solution times compared to BFS, they generated fewer nodes, demonstrating efficient memory usage. The limits of this algorithm should also be examined in a generalized rush hour board larger than 6x6.

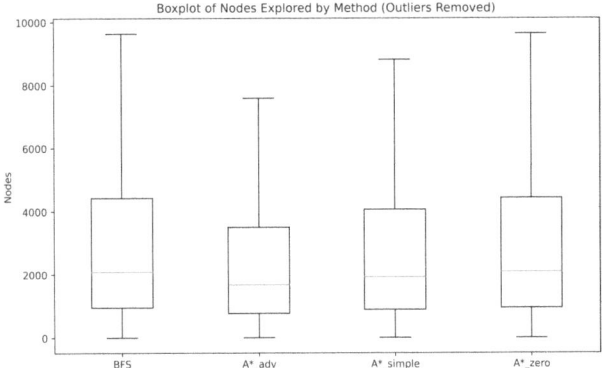

Fig. 8. Boxplot of nodes explored by method (outliers removed).

In contrast, A* zero explored a marginally higher number of nodes compared to BFS, but exhibited substantially slower performance, suggesting possible inefficiencies in this variant of the A* algorithm. Given that the A* algorithm with a zero heuristic should theoretically behave similarly to a BFS implementation [9], the significantly longer solution time warrants investigation into the implementation specifics of the code. A plausible explanation could be the utilization of the priority queue. However, the A* implementation was used primarily for benchmarking heuristics, which will subsequently serve for

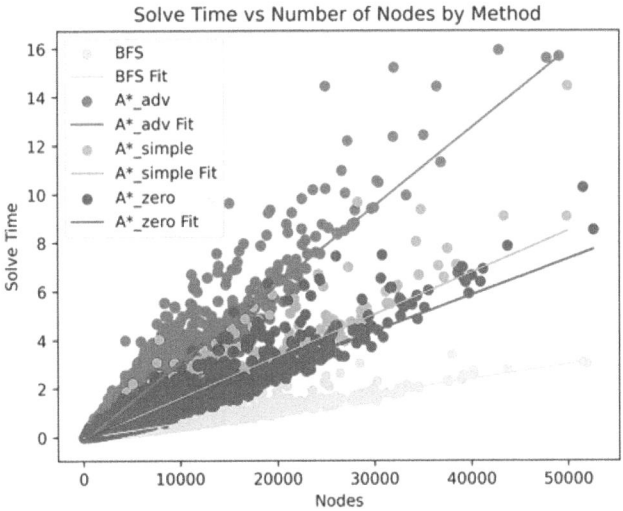

Fig. 9. Solution time vs number of nodes by method.

training neural networks. Therefore, this inefficiency does not affect the final solution. We, thus, opted against further code optimization of this approach.

Table 1 displays specific average solve times and the number of nodes generated. All methods showed identical average differences in the objective values of an optimal solution, indicating that the algorithms did not differ in the quality of obtained solutions. It is important to note that the optimal number of moves in the source database considers a car's move across multiple blocks as a single move, whereas this solution counts every move individually, leading to a non-zero difference. Therefore, a nonzero deviation in the objective value can still correspond to an optimal solution.

Table 1. Average solution time, number of nodes, and deviation to optimal solution for the considered methods.

method	Avg. Solve time	Avg. Nodes	Δ
A* adv	0.745444	2823.4933	-15.9641
A* simple	0.440894	3222.8746	-15.9641
A* zero	0.414143	3516.2348	-15.9641
BFS	0.186199	3498.8122	-15.9641

5 Conclusion and Future Work

The BFS implementation has been demonstrated to be efficient and will serve as a solid foundation for the upcoming BF-Ksubs implementation. The A* implementation, along with its heuristics, can be employed for two main purposes: training the necessary transformer networks for subgoal functions and providing a baseline benchmark for evaluation. A future area of interest would be to generalize the implementation from a 6x6 board to an n*m board. This would provide an understanding of the limitations of BFS, A*, and eventually, the Subgoal Search implementation. The code described in this paper is available at https://github.com/TipsyPanda/subgoal-search.

Disclosure of Interests. The authors have no competing interests to declare that are relevant to the content of this article.

References

1. Brunner, J., Chung, L., Demaine, E.D., Hendrickson, D., Hesterberg, A., Suhl, A., Zeff, A.: 1 x 1 Rush Hour with Fixed Blocks is PSPACE-complete. arXiv preprint arXiv:2003.09914 (2020)
2. Cian, L., Dreossi, T., Dovier, A.: Modeling and solving the rush hour puzzle. In: CILC2022:37th Italian Conference on Computational Logic, June 29–July 1,2022, Bologna, Italy, pp. 294–306, CEUR Workshop Proceedings (2022)

3. Czechowski, K., et al.: Subgoal search for complex reasoning tasks. arXiv:2108.11204. Available: http://arxiv.org/abs/2108.11204 (2021)
4. Fogleman, M.: How I created a database of all interesting Rush Hour configurations (2018). https://www.michaelfogleman.com/rush/#DatabaseDownload. Accessed 10 July 2023
5. Hearn, R. A., Demaine, E. D.: PSPACE-completeness of sliding-block puzzles and other problems through the nondeterministic constraint logic model of computation. Theor. Comput. Sci. **343**(1), 72–96 (2005). https://www.sciencedirect.com/science/article/pii/S0304397505003105
6. Hauptman, A., Elyasaf, A., Sipper, M., Karmon, A.: GP-rush: using genetic programming to evolve solvers for the rush hour puzzle. In: Proceedings of the 11th Annual conference on Genetic and evolutionary computation, ser. GECCO '09, pp. 955–962, Association for Computing Machinery, New York, NY, USA (2009). https://doi.org/10.1145/1569901.1570032
7. Rush Hour Puzzle, https://shorturl.at/zTW36. Accessed 10 July 2023
8. Russell, S. J.: Artificial intelligence a modern approach. Pearson Education, Inc. (2010)
9. Red Blob Games: Implementation of A* (2016). https://www.redblobgames.com/pathfinding/a-star/implementation.html. Accessed 10 July 2023
10. Vries, K. L. d.: Programming Puzzle: Rush Hour Traffic Jam. https://medium.com/swlh/programming-puzzle-rush-hour-traffic-jam-3ee513e6c4ab. Accessed 23 Feb 2023

Advancing Precision Agriculture: Machine Learning-Based Crop Recommendation for Optimal Yield

Mohamed Bouni[1](✉) , Badr Hssina[1], Khadija Douzi[1], and Samira Douzi[2]

[1] Laboratory LIM, IT Department FST Mohammedia, Hassan II University, Casablanca, Morocco
medbouni@gmail.com, {badr.hssina,khadija.douzi}@fstm.ma
[2] FMPR, Mohammed V University in Rabat, Rabat, Morocco
s.douzi@um5r.ma

Abstract. In the modern agricultural landscape, it's essential for farmers to adopt advanced agricultural practices to navigate contemporary challenges such as uncontrollable costs due to supply-demand imbalances, water shortages, and climatic uncertainties. Agricultural production faces numerous obstacles, including unpredictable climate changes, inadequate irrigation systems, declining soil fertility, and conventional farming methods. Accurate crop yield forecasting is crucial for enhancing agricultural practices, offering surveillance under diverse climatic conditions, and safeguarding crop yields against various weather-related issues. Machine learning emerges as a potent tool in agriculture, utilized for predicting crop recommendations. This study employs machine learning techniques to recommend a wide range of crops, including rice, maize, jute, cotton, coconut, papaya, orange, apple, muskmelon, watermelon, grapes, mango, banana, pomegranate, lentil, black gram, mung bean, moth beans, pigeon peas, kidney beans, chickpea, and coffee, all influenced by variables such as nitrogen, phosphorus, potassium, rainfall, temperature, humidity, and pH levels. Our proposed method involves the deployment of a Fully Connected Deep Neural Network (FC-DNN) based crop recommendation system, which is then compared against various machine learning models, such as decision tree classifier, logistic regression, gradient boosting classifier, random forest classifier, K-Neighbors classifier, support vector machine classifiers, and Naïve Bayes classification, through performance comparisons evaluating their predictive accuracy.

Keywords: Precision Agriculture · Decision Tree · Gradient Boosting · Naïve Bayes · Support Vector Machine · Logistic Regression · Random Forest · Machine Learning · Fully Connected Deep Learning

1 Introduction

Predicting seed yield in diverse and unfamiliar environments poses a formidable challenge in plant breeding, requiring comprehensive datasets that capture a wide range of conditions [1]. The conventional approach to generating such datasets is hindered by the considerable time and cost involved, particularly in managing extensive plots and considering long-term environmental implications. In response to these challenges, our work introduces a novel perspective by emphasizing the integration of non-genetic and genetic characteristics in plant growth techniques, thereby facilitating a more nuanced understanding of the factors influencing seed yield [2, 3].

Our research recognizes the pivotal role of the intersection between non-genetic and genetic traits across various crop species, providing a fresh and comprehensive approach to addressing future agricultural production challenges [4, 5]. By investigating the intricate interplay of these characteristics, we aim to uncover novel insights that contribute to the development of cultivars with enhanced climatic robustness. This approach goes beyond traditional breeding methods, offering a more nuanced understanding of the factors influencing seed production under different environmental conditions, including variations in rainfall events, temperature, and other critical weather parameters [6].

2 Related Work

Weather variables' temporal variability is just as essential as their geographical variability, but it's less well known in crop production studies [7, 8]. It's essential to understand how weather unpredictability affects agricultural production in the form of universal weather modification, particularly as the frequency of harsh weather events increases. As a result, forecasting the outcomes of fluctuating surroundings on production can aid in producing smart crop breeding selections, market strategy, efficient production, and comparing results across time [9]. Crop development algorithms are traditionally presented to evaluate and foresee agricultural yield in a variety of circumstances that include climate, soil conditions, and genotype and managing features [10]. These offer a suitable description of bio-physical processes and reactions, but they contained flaws in terms of estimating input parameters and predicting outcomes in complicated and unforeseen situations [11, 12].

In previous studies, crop algorithms constructed by evaluating reaction in a small number of lines while altering a particular environmental factor used to forecast yield across environments, restricting the range of outcome [13]. To skip the hindrance of crop production algorithms, linear representations have been successfully employed to forecast production [14]. These low-capacity algorithms, on the other hand, often depends on a lesser number of variables, unable to reflect the complexities of biological cooperation and additional position specific variables.

For time series forecasting difficulties, traditional linear approaches like Autoregressive Integrated Moving Average have already been utilized [15], but these approaches are only useful to predict future actions in the identical time-series. Taking into account the significance of weather harsh conditions in crop forecasting, randomly a forest model was developed to estimate yield grid-cell variations [16]. Deep neural networks are capable of approximating any non-linear functions and are resilient to noisy data in time series prediction applications. In the face of such complicated data, such as weather parameters, growth groups and regions, and genetic data, deep neural networks would provide solutions. These algorithms are capable of learning non-linear correlations between multidimensional raw datasets (development cluster, weather variables, and group info) and anticipated yields in a very efficient manner [17].

However, employing a time series technique to combine weather factors is still necessary. Additional to precise estimation, domain specialists in the field of plant breeding can benefit greatly from the capacity to comprehend domain-related estimation outputs from a machine learning algorithm [18].

3 Motivation and Contribution

An existing challenge in forecasting crop growth suitability lies in the tendency to consider only one factor, either weather or soil, without a holistic examination of all pertinent components. Achieving the most accurate and comprehensive forecasts requires a simultaneous analysis of various elements. While a specific soil type may be optimal for a particular crop, adverse climate conditions in the region can impact production adversely. To address these limitations, our developed Crop Recommendation system takes into account a comprehensive set of variables, including rainfall, location, soil nutrients, and temperature, to assess crop suitability. The primary focus of this system is to provide farmers with crop recommendations through algorithmic processes. The integrated parameters offer users a straightforward and reliable approach to make informed decisions and plan crops effectively. The ultimate goal is to establish a robust model capable of consistently predicting crop sustainability across different states based on the interplay of soil nutrients and weather conditions.

4 Algorithms for Machine Learning

4.1 Regression Analysis: Logistic Method

Logistic Regression, a predictive modeling tool widely used in agriculture, holds significance for crop prediction by analyzing binary dependent variables and capturing complex relationships between environmental factors and crop outcomes. Its application in modeling the probability of crop success based on factors like rainfall and temperature makes it a valuable asset for farmers. To assess its effectiveness, an exploration into logistic regression's performance metrics in crop prediction becomes essential. This analysis enhances the algorithm's relevance, emphasizing its practicality in optimizing crop management and promoting agricultural sustainability.

4.2 Classification via Decision Trees

Decision Tree Classification is a powerful algorithm for supervised learning designed primarily for categorization challenges, making it highly relevant for crop prediction. Its tree-like structure, where nodes represent dataset features and branches depict decision rules, allows for systematic dataset division. Starting from the root, the algorithm selects ideal attributes, breaks down nodes, and builds decision tree nodes until the final leaf nodes are reached. In crop prediction, this method aids in categorization and regression tasks, facilitating the division of datasets into homogeneous subcategories. A more detailed exploration of its practical applications and performance in crop prediction could enhance its effectiveness in providing decision support in agriculture.

4.3 Gradient Boosting Classification

Gradient Boosting Classification, a versatile technique in machine learning, is highly effective in addressing both regression and classification challenges by mitigating bias errors. Its application in crop prediction proves influential for predicting continuous and categorizing target variables. The method systematically tackles variance and bias errors, significantly enhancing the precision of crop prediction models. Utilizing mean square error for regressors and log loss for classifiers, Gradient Boosting Classification operates on input values and targets, reconstructing the relationship between features and the target variable. Its pivotal role in minimizing errors and optimizing model performance makes it a valuable asset in agricultural decision support systems, contributing to enhanced accuracy and reliability.

4.4 K-neighbors Classification

K-Neighbors Classification, a simple yet effective method in crop prediction, operates without requiring prior knowledge of data distribution. It leverages a training dataset to determine the k-nearest neighbors using Euclidean distance, enabling accurate categorization of crops based on majority voting from neighboring data points. This approach is particularly useful when precise parametric approximations of probability density functions are unavailable. Despite its simplicity compared to some discriminant analysis techniques, K-Neighbors Classification proves valuable in predicting crop categories, showcasing adaptability and straightforward implementation. Its ability to handle diverse datasets makes it a relevant and practical tool for predicting crop types based on spatial characteristics.

4.5 Random Forest Classification

Random Forest Classification, a collaborative learning system, proves highly effective in crop prediction by constructing multiple decision trees during training and aggregating their outputs to determine the mean estimation. This algorithm addresses overfitting concerns encountered in decision tree models, enhancing generalizability in agricultural applications. Through a combination of bagging and random feature selection, it mitigates overfitting risks and excels at handling three-dimensional data relevant to crop parameters. The method's versatility in providing accurate recommendations contributes to its significance in precision agriculture, offering practical solutions for optimizing crop yields and promoting sustainability in agricultural practices.

4.6 Naïve Bayes Classification

Naïve Bayes Classification, rooted in Bayes' theorem, is applied effectively in crop prediction within agriculture, leveraging its probabilistic nature. Tailored for scenarios where feature independence prevails, it efficiently handles diverse datasets with various crop-related parameters. By assessing variable impacts independently on specific crop groups, the algorithm predicts crop categories with a focus on conditional independence. Despite simplifying assumptions, research demonstrates competitive performance against other methods, affirming its relevance in crop recommendation systems.

Its application to extensive databases highlights its significance, providing a valuable tool for decision-making in precision agriculture and sustainable farming.

4.7 Classification with Support Vector Machines

Support Vector Machine (SVM) Classification is a powerful supervised learning algorithm widely applicable in crop prediction. It excels in categorizing and classifying crops based on diverse features, constructing a discriminative hyperplane in multidimensional space to delineate clear boundaries between different crop classes. In the context of agriculture, SVM proves valuable for making informed decisions about crop selection, especially in intricate datasets with multiple variables. Its versatility extends to handling the complexity of agricultural data, contributing to precision agriculture. SVM's role in classification, regression, and outlier detection enhances its utility, providing farmers with a tool to optimize crop yields by predicting suitable crops based on environmental factors. The algorithm's robust performance in various agricultural settings underscores its effectiveness, establishing it as a valuable asset in crop prediction models.

4.8 Proposed Fully Connected Deep Neural Network (FC-DNN)

The Deep Neural Network with Full Connectivity (FC-DNN) acts as a powerful instrument for forecasting crop yields, addressing the complexities of diverse agricultural variables such as rainfall, soil nutrients, and temperature. This artificial neural network, characterized by interconnected nodes across layers, proves instrumental in delivering precise crop recommendations. Despite inherent challenges related to processing power and the risk of overfitting, FC-DNN excels in capturing intricate relationships within agricultural datasets, contributing to the accuracy of crop predictions. The network's architecture involves five fully-connected layers, and the activation functions, including hyperbolic tangent (The hyperbolic tangent function, tanh, is frequently employed in the concealed layers of neural networks, aiding the network in capturing intricate relationships within the data. This characteristic makes tanh a preferred choice for enhancing the network's capability to learn and represent complex patterns in the hidden layers) and modified softmax, enhance the classification of different crop types. The output manifests as a probability vector, providing a nuanced assessment of each recommended crop's likelihood. This approach, depicted in Fig. 1, holds promise for revolutionizing agricultural decision-making, offering a comprehensive framework for optimizing crop choices and increasing overall productivity.

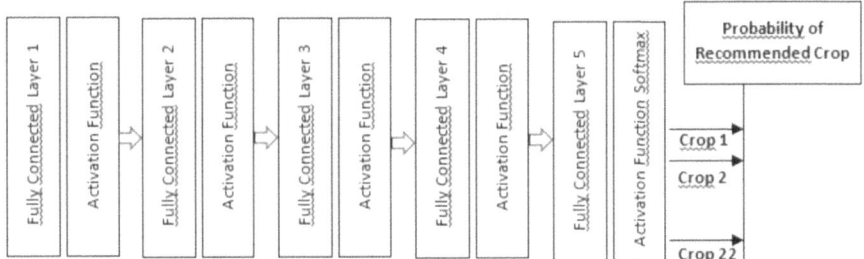

Fig. 1. Architecture of the Fully Connected Deep Neural Network for Crop Recommendation

5 Implementation Methodologies

The implementation steps of proposed approach are shown in Fig. 2

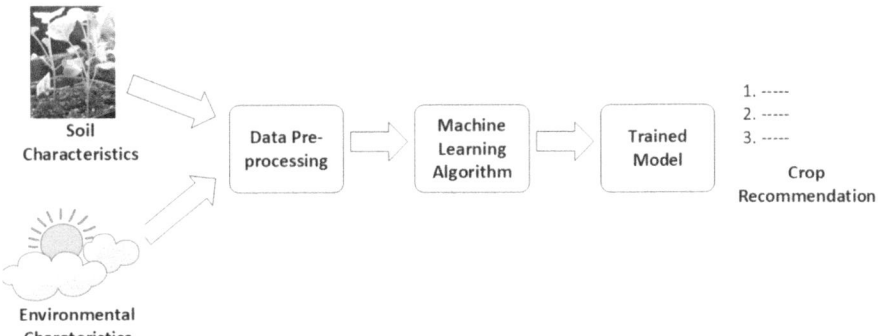

Fig. 2. Proposed Approach for ML Crop Recommendation

5.1 Dataset Collection

During the implementation phase, a critical task is the comprehensive examination of the dataset to uncover potential correlations among its various attributes. The success of the machine learning approach depends on the volume and integrity of the parameters as well as the training dataset's quality. This research thoroughly investigated several datasets, choosing configurations meticulously to maximize outcomes. The data, obtained from Kaggle, are detailed in Tables 1 and 2, providing an extensive view of the datasets used in this study. The Crop Recommendation dataset, which is fundamental to model training, includes vital factors such as humidity, temperature, mean rainfall, soil pH, nitrogen, potassium, and phosphorus ratios, crucial for predicting yields. Following predictions, additional datasets like Soil Types and Crop Types are employed to classify and pinpoint the expected crop yields. Many studies have looked at environmental factors to determine agricultural sustainability; some have utilized yield as a main factor, whereas others have merely looked at financial factors. We wanted to combine meteorological factors like temperature, ph with soil data like soil nutrients and rainfall to provide the farmer a reliable and precise recommendation for which crop to plant on his land. The read function is used to load the dataset.

Table 1. Dataset details of crop recommendation

Attributes	Values
Source	Crop Recommendation
Number of Samples	2200
Attributes	8
Usage	Classification
Label Count	22

Table 2. Description of recommendation parameters

Variable Attributes	Description Values
N	Ratio of Nitrogen content in soil
P	Ratio of Phosphorous content in soil
K	Ratio of Potassium content in soil 8
temperature	Temperature in degree Celsius
humidity	Relative humidity in %22
ph	Ph value of the soil
rainfall	Rainfall in mm
label	Various types of Crops

5.2 Data Preprocessing

Real-world data can have missing values, and noise while in an unsuitable format, making it impossible to utilize in machine learning methods directly. Preprocessing data is important for cleaning and preparing data for such algorithms, which enhances the model's efficiency and accuracy. Preprocessing data is critical since it cleans and prepares data for use in machine learning algorithms. Preprocessing aims to remove any abnormalities or erroneous dataset, as well as filling in any gaps in the data. There are two methods for resolving missing data. The first approach is to remove the entire row that contains the inaccurate or missing data. Although this strategy is simple to apply, it works best with huge datasets. When applied to small datasets, this method can result in significant data reduction, particularly when there are numerous missing values. This will have a big impact on how accurate the result is. Since our dataset is small, we will not employ the method of deleting full row. Instead, in the crop recommendation dataset from Kaggle website contains raw data whose missing entries are filled (cleaning). Additionally, for normalization we use feature scaling in the dataset. When the input from users is in the form of string, then the data is transformed using the technique where we convert string to numeric data.

5.3 Training and Testing Data

To obtain credible results, we employed machine learning techniques to both train and test the proposed system across diverse scenarios. In this process, the data was trained to predict the crop yield based on various input features, encompassing soil nutrients and environmental conditions. We provided a diverse set of input features and trained the dataset to predict the specific crop yield associated with them. Using X test data, the dataset was fitted to the X, Y trained values, and predictions were generated. The model underwent training for 100 epochs, and the optimal model, characterized by the lowest loss, was selected for subsequent evaluation and testing. The data flow of the proposed algorithm is visually depicted in Fig. 3

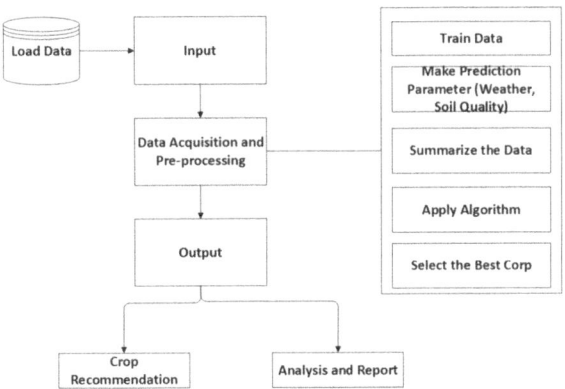

Fig. 3. Proposed Approach for Data-Driven Crop Recommendation Flow

5.4 Implementation Theory

Researchers are increasingly committing their resources and time to assisting in the solution of the problems in agriculture as more people become aware of it. Normalized Greedy Forest has been used to identify an appropriate crop sequence for a specific time stamp in various works. Another method employs historical atmospheric datasets as a training set for a model. The system has been programmed to detect weather patterns that are bad for apple production. Based on weather patterns, the production of crops would then be correctly forecast. Several strategies, like Neural Networks, K-Nearest Neighbors, and Greedy Forest, etc., are utilized to pick a crop based on the expected production rate, that is influenced by numerous criteria. The system also includes pesticide planning and online exchange based on agricultural commodities. Each study's authors focused on a single variable (weather or soil) for forecasting crop growth appropriateness, which we discovered to be a fault in all of these key published studies. However, the greatest and most accurate forecast, each of these components should be examined simultaneously. A particular soil type may be ideal for growing a specific crop, the production may suffer if the climate circumstances in the location unfavorable for that crop.

6 Results and Discussion

The anticipated results are gauged using the accuracy metric, which evaluates how closely a measurement aligns with its true value. Precision is calculated by the ratio of the sum of true negatives and true positives to the total of true negatives, true positives, false positives, and false negatives. The most effective model is identified by its highest level of accuracy. We apply a variety of algorithms to suggest a range of crops, such as rice, maize, jute, cotton, coconut, papaya, orange, apple, muskmelon, watermelon, grapes, mango, banana, pomegranate, lentil, black gram, mung bean, moth beans, pigeon peas, kidney beans, chickpea, and coffee. These crops are influenced by factors like nitrogen, phosphorus, potassium, rainfall, temperature, humidity, and pH levels. The interrelations between features impacting crop yield are depicted in Fig. 4. The correlation matrix provided reveals relationships between various agricultural factors. Notably, Nitrogen (N) and Phosphorus (P) exhibit a moderate inverse correlation (-0.231460), suggesting a tendency for one to increase as the other decreases. In contrast, Phosphorus (P) and Potassium (K) display a strong positive correlation (0.736232), indicating a robust positive association between these two elements. Potassium (K) and temperature demonstrate a moderate inverse correlation (-0.160387), suggesting that as temperature increases, Potassium levels tend to decrease. A weak positive correlation (0.094423) is observed between humidity and pH, implying a subtle tendency for higher humidity to correspond with a slightly higher pH. Additionally, temperature and rainfall display a moderate positive correlation (0.205320), hinting at a tendency for increased rainfall to coincide with higher temperatures.

Whereas, the overall/relative feature importance for each factor effecting the crop is given in Fig. 5 which shows that nitrogen emerges as the most crucial factor with a feature importance of nearly 0.11, signifying its substantial impact on the model's output. Following closely is phosphorus, exhibiting a feature importance of 0.151, underscoring its noteworthy contribution to the model's decision-making process. Potassium holds a significant position with a feature importance of 0.18, indicating its pronounced effect on the model's predictive accuracy. Furthermore, environmental factors such as humidity and rainfall stand out prominently, boasting feature importance values of 0.22 and 0.24, respectively. These values emphasize the considerable influence that fluctuations in humidity and rainfall levels wield over the model's predictions. On the other hand, temperature and pH, while still significant, exhibit comparatively lower feature importance values of around 0.075 and 0.055.

In Fig. 6, the confusion matrix is utilized to assess the performance of the classification model, offering a precise breakdown of both accurate and inaccurate predictions. Rows in this matrix align with the actual classes, and columns correspond to the classes as predicted by the model. Diagonal entries represent correct predictions for each class, whereas entries off the diagonal reflect misclassifications. This confusion matrix reveals the model's generally robust performance, as indicated by high values along the diagonal. Specifically, class 8 (pH) shows a rare misclassification, with one instance mistakenly identified as class 20 (rainfall). In a similar vein, class 20 (rainfall) sees three instances wrongly classified as class 8 (pH).

Table 3 presents the performance indicators for various machine learning models, scrutinizing accuracy, recall, F1 score. Of particular note is the Fully Connected Deep Neural Network (FC-DNN), which stands out for its exceptional accuracy rate of 99.51%, showcasing its precise predictive ability. The Random Forest Classifier also delivers a solid performance, with an accuracy of 98.91%, closely followed by the Decision Tree Classifier at 97.31%. These models consistently achieve high recall and F1 scores, highlighting their effectiveness in correctly identifying relevant instances while maintaining a balance between precision and recall. Although the Logistic Regression model proves to be reliable, it scores a slightly lower accuracy of 91.51%. Together, these performance metrics offer a thorough evaluation of the models' strengths, guiding the choice of the most appropriate model based on specific needs, whether it be reducing errors or enhancing accuracy.

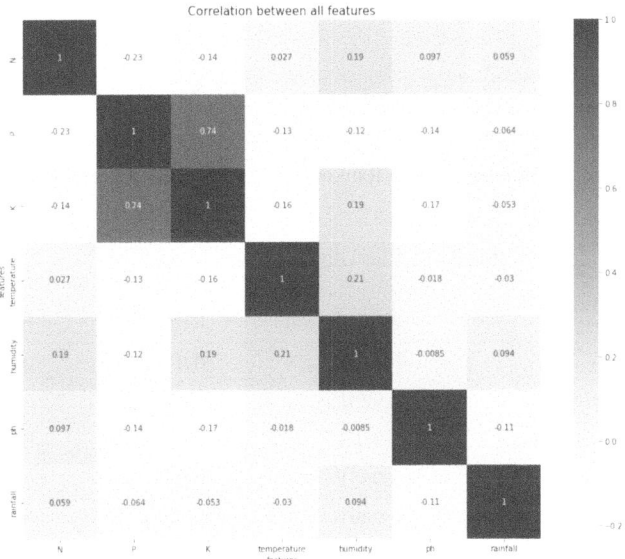

Fig. 4. Correlation matrix for variable interdependency heatmap in crop analysis

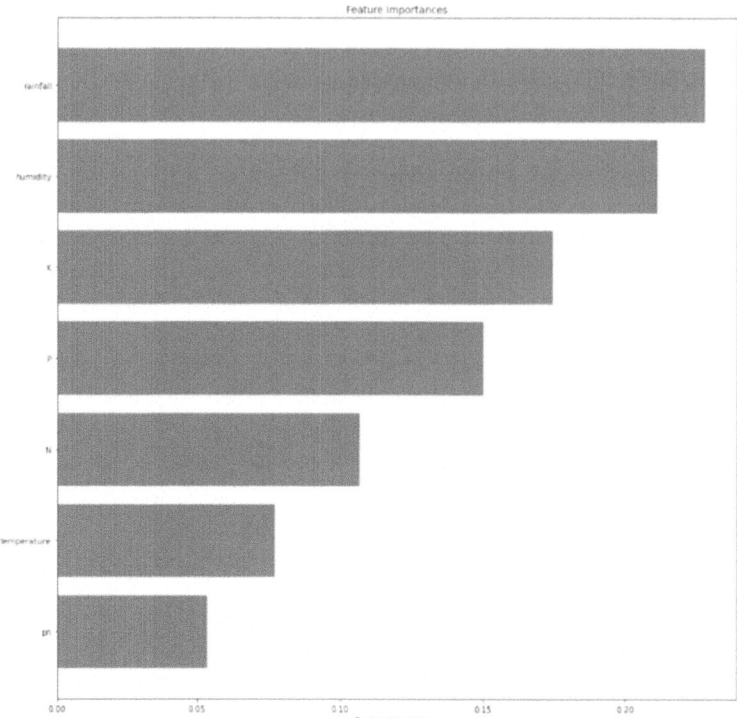

Fig. 5. Feature Importance Rankings in Crop Yield Prediction Model

Table 3. Metrics comparison of the models

Name of model	Accuracy	Recall	F1
logistic regression	0.9151	0.9151	0.9151
decision tree classifier	0.9731	0.9731	0.9731
gradient boosting classifier	0.9663	0.9663	0.9663
K-Neighbor classifier	0.9551	0.9551	0.9551
random forest classifier	0.9891	0.9891	0.9891
Naïve Bayes	0.9754	0.9754	0.9754
SVM	0.9595	0.9595	0.9595
FC-DNN	0.9951	0.9951	0.9951

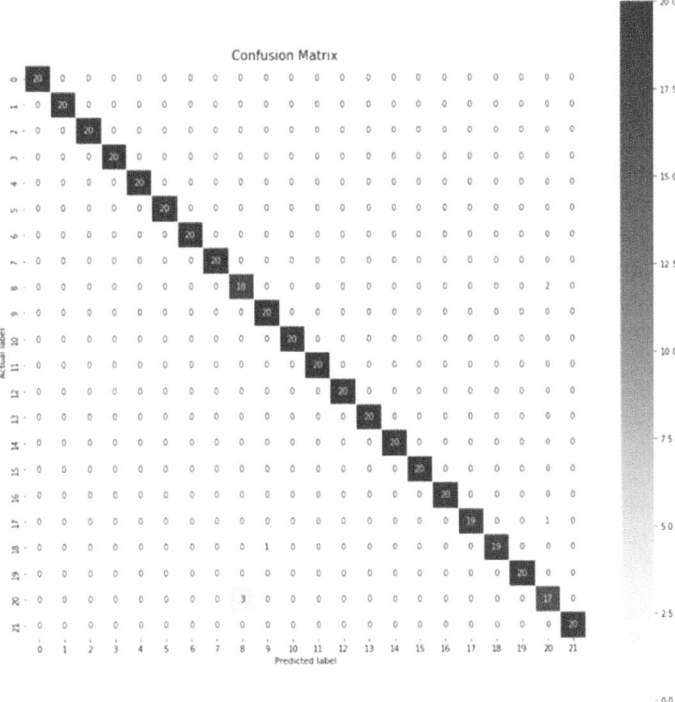

Fig. 6. Confusion Matrix for Crop Prediction Using FC-DNN Model

7 Conclusion

Machine learning technology serves as a pivotal resource for farmers, providing crucial data that may be beyond their current knowledge, thus enhancing crop yields and streamlining operational efficiency. Additionally, it plays a significant role in minimizing economic risks associated with farming. With the potential for internet-based expansion, this innovative method has the capacity to support a multitude of farmers across various regions. The system's precision exemplifies its utility, offering substantial accuracy improvements. Future developments include coupling the crop recommendation engine with a yield prediction component, offering farmers projections of harvest volumes for their chosen crops.

In conclusion, our research underscores the transformative role of machine learning in agriculture, particularly in the realm of crop recommendation systems. The significance of this technology lies in its potential to bridge knowledge gaps among farmers, reducing yield loss, increasing productivity, and safeguarding against financial setbacks. As we envision the broader implementation of this technique, its scalability to the online platform holds the promise of reaching millions of farmers nationwide, offering practical solutions to enhance the efficiency and sustainability of farming practices.

The demonstrated efficacy of our proposed method speaks to its potential to deliver a higher level of accuracy in crop recommendation, marking a substantial advancement in precision agriculture. However, the conclusion also acknowledges the need for further justification and refinement, especially regarding the selection of activation functions in the deep neural network. Looking ahead, the integration of our crop recommendation system with a complementary crop estimator holds exciting prospects. This synergy will not only guide farmers in selecting the most suitable crops but will also empower them with production estimates, creating a more holistic approach to decision-making.

Our research stands at the forefront of leveraging machine learning for the betterment of agriculture, offering practical solutions and paving the way for ongoing research and development in this dynamic and promising field. The commitment is to continually enhance the accuracy, accessibility, and applicability of our technology, fostering a more resilient and productive agricultural landscape.

References

1. Nti, I.K., Zaman, A., Nyarko-Boateng, O., Adekoya, A.F., Keyeremeh, A.F.: A predictive analytics model for crop suitability and productivity with tree-based ensemble learning. Decis. Anal. J. **8**, 100311 (2023). https://doi.org/10.1016/j.dajour.2023.100311
2. Iniyan, S., Varma, V.A., Naidu, C.T.: Crop yield prediction using machine learning techniques. Adv. Eng. Softw. **175**, 103326 (2023). https://doi.org/10.1016/j.advengsoft.2022.103326.
3. Tamil Selvi, M., Jaison, B.: Lemuria: a novel future crop prediction algorithm using data mining. Comput. J. **65**(3), 655–666 (2022). https://doi.org/10.1093/comjnl/bxaa093
4. Vogel, E., et al.: The effects of climate extremes on global agricultural yields. Environ. Res. Lett. **14**(5), 054010 (2019). https:// doi.org/https://doi.org/10.1088/1748-9326/ab154b
5. Pawar, M. Chillarge, G.: Soil toxicity prediction and recommendation system using data mining In: Precision Agriculture, 3rd International Conference for Convergence in Technology (I2CT),2018, pp. 1–5 (2018). https://doi.org/10.1109/I2CT.2018.8529754
6. Arooj, A., Riaz, M., Akram, M.N.: Evaluation of predictive data mining algorithms in soil data classification for optimized crop recommendation. In: 2018 International Conference on Advancements in Computational Sciences (ICACS), pp. 1–6 (2018). https://doi.org/10.1109/ICACS.2018.8333275.
7. Patel, K., Patel, H.B.: A state-of-the-art survey on recommendation system and prospective extensions. Comput. Electron. Agric. **178**, 105779, ISSN 0168–1699 (2020). https://doi.org/10.1016/j.compag.2020.105779
8. Chaudhari, A., Beldar, M., Dichwalkar, R., Dholay, S.: Crop Recommendation and its Optimal Pricing using ShopBot. In: 2020 International Conference on Smart Electronics and Communication (ICOSEC), pp. 36–41 (2020). https://doi.org/10.1109/ICOSEC49089.2020.9215411
9. Pudumalar, S., Ramanujam, E., Rajashree, R.H., Kavya, C., Kiruthika, T., Nisha, J.: Crop recommendation system for precision agriculture. In: 2016 Eighth International Conference on Advanced Computing (ICoAC), pp. 32–36 (2017). https://doi.org/10.1109/ICoAC.2017.7951740.
10. Anguraj, K., Thiyaneswaran, B., Megashree, G., Shri, J.G.P., Navya, S., Jayanthi, J.: Crop recommendation on analyzing soil using machine learning. Turk. J. Comput. Math. Edu. **12**(6), 1784–1791 (2021). https://www.proquest.com/scholarly-journals/crop-recommendation-on-analyzing-soil-using/docview/2623930939/se-2
11. Sharma, A., Bhargava, M., Khanna, A.V.: AI-Farm: a crop recommendation system. In: 2021 International Conference on Advances in Computing and Communications (ICACC), pp. 1–7 (2021). https://doi.org/10.1109/ICACC-202152719.2021.9708104

12. Dighe, D.: Survey of crop recommendation system. Int. Res. J. Eng. Technol. (IRJET) e-ISSN: 05(11), 2395–0056 (2018)
13. Esquerdo, J.C., Zullo Júnior, J., Antunes, J.F.: Use of NDVI/AVHRR time-series profiles for soybean crop monitoring in Brazil. Int. J. Remote Sens. **32**(13), 3711–3727, https://doi.org/10.1080/01431161003764112. (2011)
14. Sodha, D., Saha, G.: Crop management of agricultural products using time series analysis. In: 2016 IEEE International Conference on Recent Trends in Electronics, Information & Communication Technology (RTEICT), pp. 1456–1460 (2016). https://doi.org/10.1109/RTEICT.2016.7808073.
15. Priya, R., Ramesh, D., Khosla, E.: Crop prediction on the region belts of India: a naïve bayes mapreduce precision agricultural model. In: 2018 International Conference on Advances in Computing, Communications and Informatics (ICACCI), pp. 99–104 (2018). https://doi.org/10.1109/ICACCI.2018.8554948.
16. Mokarrama, M.J., Arefin, M.S.: RSF: a recommendation system for farmers. In: 2017 IEEE Region 10 Humanitarian Technology Conference (R10-HTC), pp. 843–850 (2017). https://doi.org/10.1109/R10-HTC.2017.8289086.
17. Raja, S.K.S., Rishi, R., Sundaresan, E., Srijit, V.: Demand based crop recommender system for farmers. In: 2017 IEEE Technological Innovations in ICT for Agriculture and Rural Development (TIAR), pp. 194–199 (2017). https://doi.org/10.1109/TIAR.2017.8273714
18. Katarya, R., Raturi, A., Mehndiratta, A., Thapper, A.: Impact of machine learning techniques in precision agriculture. In: 2020 3rd International Conference on Emerging Technologies in Computer Engineering: Machine Learning and Internet of Things (ICETCE), pp. 1–6 (2020). https://doi.org/10.1109/ICETCE48199.2020.9091741.

Pattern Recognition

Low-Cost Biodegradable Composite-Shaped Split Ring Resonator (SRR) Based Sensor for Breast Cancer Assessment

Madan Kumar Sharma[✉], Degala Satyanarayana, and Abdullah Said Alkalbani

College of Engineering, University of Buraimi, Al Buraimi, Oman
madan.k@uob.edu.om

Abstract. This study introduces a biodegradable sensor with a composite-SRR design for breast tumor analysis. The multi-resonance behavior of the sensor is achieved by etching the SRR structure in the sensor elements. Resonate frequency is tuned by inductive and capacitive nature of the rectangular and circular shaped sensing elements, which is known as hybrid shaped structure of the sensor. The parametric study has resulted in the development of a low-cost, flexible, and readily biodegradable Cellulose Nanofibril (CNF) substrate measuring 30 mm × 100 mm. The proposed sensor is offered the multi-resonating behavior with sharp resonating frequency, there are total 10 resonating dips (S11 < -10 dB) at the frequency of 3.2 GHz, 6.9 GHz, 7.5 GHz, 8.09 GHz, 9.4 GHz 10.18 GHz, 16.3 GHz, 17.5 GHz, 19.4 GHz, and 20.3 GHz are offered by the sensor. Breast models, both normal and malignant, are generated within the biomedical exposure environment of Computer Simulation Technology (CST). Further, these models are tested with proposed sensor for tumor detection. Evaluated s-parameters for normal and malignant model (with 10 mm tumor) are exhibited a (10.22 dB) substantial difference in the reflection results. The proposed sensor flexibility and biodegradability compare with existing literature.

Keywords: Sensors · SRR · Hybrid-shaped · Tumor · Breast Model

1 Introduction

The breast cancer is a life challenging disease among the women around the globe. The disease commonly occurred above the age of 50, in the early stage there is no significant symptoms identified by the doctors, except seen lumps, red races around the nipple [1]. In this scenario, early diagnosis of the disease becoming more challenging. Nevertheless, large tumors can be evaluated using a range of screening tools, including mammograms, X-rays, ultrasound, and MRI scans. However, frequent screening with these tools may pose risks to the patient [2]. Although, MRI can offer exact location of the tumor but due to high cost it is not always accessible by every hospital. However, more accurate diagnosis is carried out using biopsy method, but sample extraction is painful for the patient and can cause fast spreading of the cancerous cell in the other organs [3]. These screening methods have shown limitations in terms of harmful ionizing radiation and unable to detect the early-stage tumors.

In the last decade, microwave sensing and imaging became an alternative early-stage diagnosis tool [4, 5]. The Microwave sensing methods are accurate, non-invasive, patient comfortable and easily accessible solution over the existing screening methods. There are several microwave sensors and setups reported in [6–16] for various medical applications. A compact slotted antenna sensor setup has been deployed in [6] for breast cancer screening, tumor was asses from s-parameter results. In [7], single feed dipole array is experimented with breast phantom for tumor detection. The sensor resonated at the 11.5 MHz narrowband frequency, S-parameters magnitude and phase were involved in the tumor analysis. The radar-based detection systems using Vivaldi-shaped antenna sensor has been utilized for tumor detection in [8], healthy and malignant cells discriminated using Complex Natural Response (CNR) method. A mm-wave antenna sensor has been reported in [9], small size (1 mm, 2mm, 3 mm) tumor analysis have been carried out by measuring deviation of S-parameters. In [10], to obtain the better dielectric contrast a waveguide method have been presented to operate in THz frequency range for breast tumor detection, but reported techniques required biopsy sample for tumor detection. A textile antenna sensor has been implemented in [11] for tumor detection, the sensor operated in the range of 2.2 GHz- 8GHz frequency. Dielectric and S-parameters were evaluated for tumor analysis. In [12], metamaterial-based array has been used to assess a tumor in the breast model. Ultra-wideband multi-element sensors have been experimented with breast phantoms in [13–15] to detect the less than 20 mm tumor. A wearable textile antenna has been developed in [16] for various wireless medical systems. The antenna was fabricated on the jeans substrate and operated in 0.9 GHz -6 GHz frequency.

Implemented sensors as discussed in the existing literature have their own advantages and limitations. Most of the sensors are developed in literature are non-biodegradable, non-disposable and non-flexible. Repeatedly use of the same sensor in the setup a concern of patient hygiene and non-flexibility of the sensor can be exhibit un-comfortable to the patient under examination. Another important aspect of tumor assessment process, is sensor sensitivity, which is largely depend upon the relative resonating behavior of the sensor while sample under test. Wide- band operating resonating behavior and single resonating behavior of the sensor can show limited sensitivity while interact with sample. To address the concerns a new multi-resonating sensor is proposed in this work. In comparison, multi-resonating behavior of the sensor is offered improved sensitivity and sufficient approach for tumor detection capability. The proposed work is contributing in terms of following major points:

1. To develop a new composite-shaped Split-Ring Resonator based multi-resonating sensor.
2. To Investigate the best suitable single used bio-degradable and flexible substrate for sensor development.
3. Test the sensor performance in terms of s-parameters, current distribution and other performance indicators.
4. To test the breast phantom using proposed sensor to detect the small size (10 mm) tumor.

The rest of the work is structured as follows: In Sect. 2, the design and development of the proposed sensor is outlined, while Sect. 3 explores into the analysis of the breast phantom using said sensor. Sect. 4 details the paper's conclusion. Additionally, Table 1 provides a comparison between this work and existing research.

Table 1. Comparison study of the proposed sensor with existing literature

Ref.	Size of the sensor (mm²)	Resonating Frequency (GHz)	Substrate Type	Flexibility	Biodegradability
[2]	123 × 30	1–1.3	RO4003C	No	No
[3]	42 × 73	5–10	-	No	No
[4]	5 × 5	32.63–33.96	RT 5880	No	No
[6]	50 × 50	2.2 - 8	Textile	Yes	No
[7]	22 × 22	2 - 12	FR4	No	No
[10]	60 × 48	0.9 - 6	Jeans	Yes	Yes
This research	100 × 30	3.2, 6.9, 7.5, 8.09, 9.4, 10.18, 16.3, 17.5, 19.4, and 20.3 (multi-resonating)	Plant based CNF	Yes	Yes

2 Design and Development of the Sensor

The sensor is made on a flexible and eco-friendly material called Cellulose Nano-fibril (CNF). It has overall size 30 mm × 100 mm, dielectric constant of 2.9 and a loss tangent of 0.03. Figure 1 shows the design and dimensions of the sensor. The sensor is comprised of three unique, split-ring shaped elements, hence the name Hybrid-shaped Split-Ring Resonator (HSRR) sensor.

The first element is a rectangular shaped in which four rectangular-shaped SRR structures are etched. Second element is a circular-shaped structure which has four

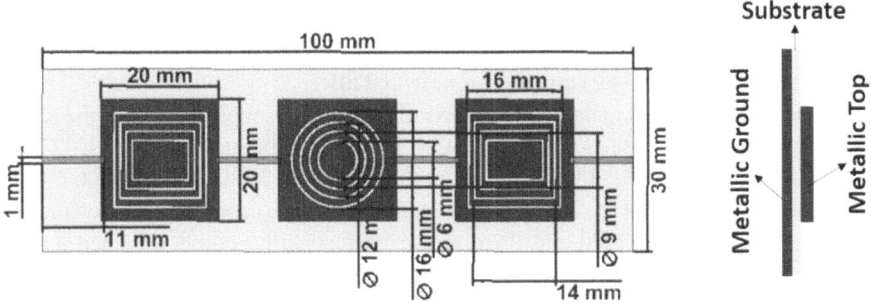

Fig. 1. Proposed HSRR sensor structure with their optimized dimensions.

circular-shaped SRR and next element is the same as first element. Figure 2 illustrates the S-parameters (reflection and transmission coefficients) of the proposed sensor. The sensor is offered the multi-resonance at the frequency 3.2 GHz, 6.9 GHz, 7.5 GHz, 8.09 GHz,9.4 GHz 10.18 GHz, 16.3 GHz, 17.5 GHz, 19.4 GHz, and 20.3 GHz. The transmission coefficient of the sensor more than -20 dB except the range of frequency 5.3 GHz-11.9 GHz.

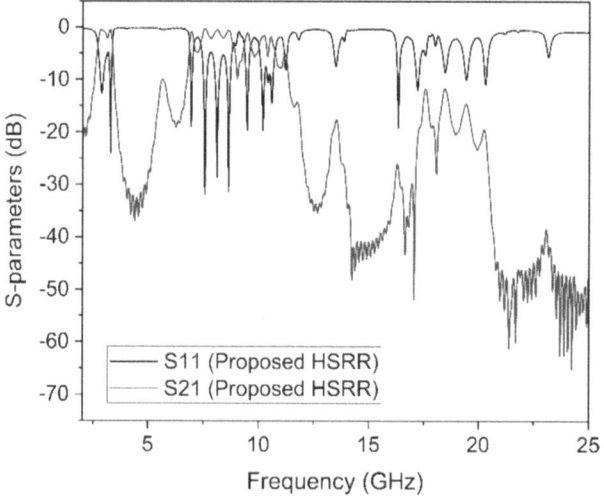

Fig. 2. Results of S-parameters for proposed HSRR sensor

2.1 Development Steps of the Sensor

Development steps of the HSRR sensor is carried out in three steps as shown in Fig. 3, initially a single element with perturbed four rectangular SRR is configured in the center of the substrate and connected both side with feedline. Reflection and transmission coefficients for each step is analyzed from Fig. 4 and Fig. 5 respectively. Reflection results of the step-1 as depicted in Fig. 4, the resonating points are -22 dB, -21 dB,18.1 dB, -13 dB, and -11 dB at 3.6 GHz,7.5 GHz,10.9 GHz,13.7 GHz, and 18 GHz respectively. However, stop band responses are approximately closer to -5dB. The transmission results for this case are observed as very poor or near to 0 dB.

In the next step, two rectangular-shaped resonators are implemented in the sensor configuration. In this case, sensor resonating points are obtained as -16 dB, -21 dB, -10 dB, -36.5 dB, -26 dB, -13 dB, -11 dB at the frequency 2.6 GHz, 3.6 GHz, 6 GHz, 7.5 GHz, 11 GHz, 14 GHz, 22.6 GHz respectively. In this case sharpness of the resonances have been improved and there are seven resonating points obtained as well as transmission coefficient also improved with -10 dB. In the final step, a circular-shaped sensing element with etched four circular-shaped SRR inserted between the two rectangular sensing elements. This configuration of the sensor is known as proposed

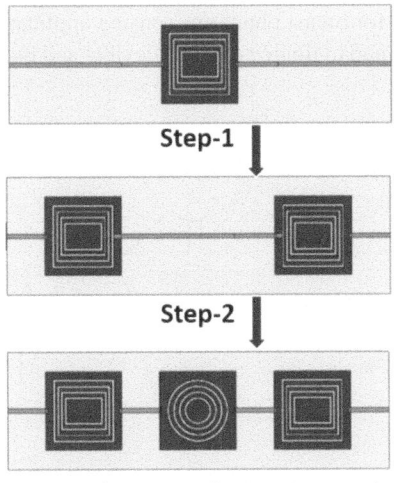

Fig. 3. Evolution process of the proposed HSRR sensor

hybrid-shaped SRR sensor. Final results are exhibited 12 resonance points with value of -22 dB, -18 dB, -24 dB, -24.5 dB, -26.8 dB, -17.24 dB, -14.8 dB, -13 dB, -14.46 dB, -19.8 dB, -20.8 dB, 20.8 dB, at the frequency 3.7 GHz, 7.8 GHz, 8.5 GHz,9.1 GHz, 9.7 GHz,10.7 GHz, 11.5 GHz, 15.2 GHz, 18.4 GHz, 19.47 GHz, 20.9 GHz, and 22 GHz respectively. Except for specific frequency ranges between 3.1 GHz and 3.7 GHz, as well as 6 GHz and 13 GHz, the transmission coefficient of the final configuration more than -20 dB. The sharp multi-resonance and improved transmission coefficient results

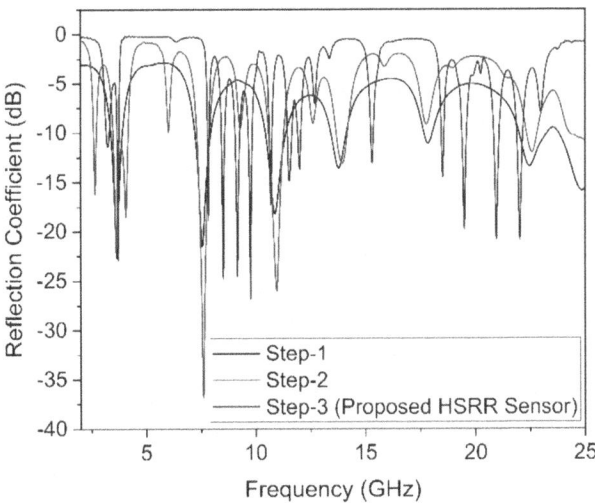

Fig. 4. Reflection results for the development steps of the HSRR sensor

of the final step are useful for breast phantoms sensing applications. In the next section, parametric study is performed to finalized better flexible and biodegradable substrate for sensor development.

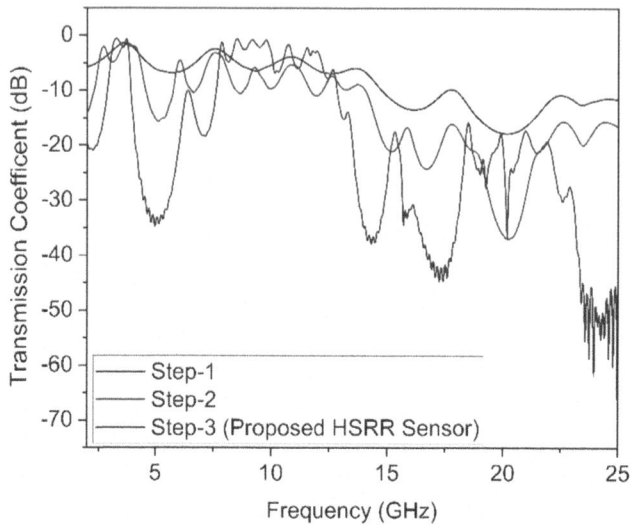

Fig. 5. Transmission coefficient results for the development steps of the HSRR sensor

2.2 Substrate Based Parametric Analysis of the Proposed HSRR Sensor

To investigate a flexible and biodegradable substrate for sensor development, the parametric analysis of the different substrates is performed in this section. There are four flexible substrates viz: wash cotton, jeans, paper and plant-based CNF with relative permittivity of 1.51, 1.55, 2.3, and 2.9 respectively are selected in the analysis. For each case S-parameters are evaluated as shown in Fig. 6. As reflection coefficient results represented in Fig. 6 (a), wash cotton and jeans substrate have approximately same resonating points with minimum magnitude of reflection -13 dB only and stop band points are poor at the frequency range 9 GHz to 14 GHz with magnitude of -6 dB. The paper and CNF substrate have shown significant improvement in the resonating points with the deep dips up to -31 dB as well as exhibits more resonating points compare to cotton and jeans substrate. The transmission coefficients are also improved comparatively. However, paper and CNF each substrate has good multi-resonating points but, in terms of flexibility and reliability and biodegradability CNF is better than paper-based substrate. Therefore, proposed sensor is implemented on the CNF substrate.

The surface current distribution as depicted in Fig. 7 is also evaluated for analyzing the losses across the other part of the sensor except maximum energy coupled across the sensing elements. The surface current distribution is evaluated with excited the port-1, at the resonating frequency 3.3 GHz, 8.5 GHz, 16.5 GHz. The evaluated results are shown that, maximum current is coupled across the sensing elements through feedline on the

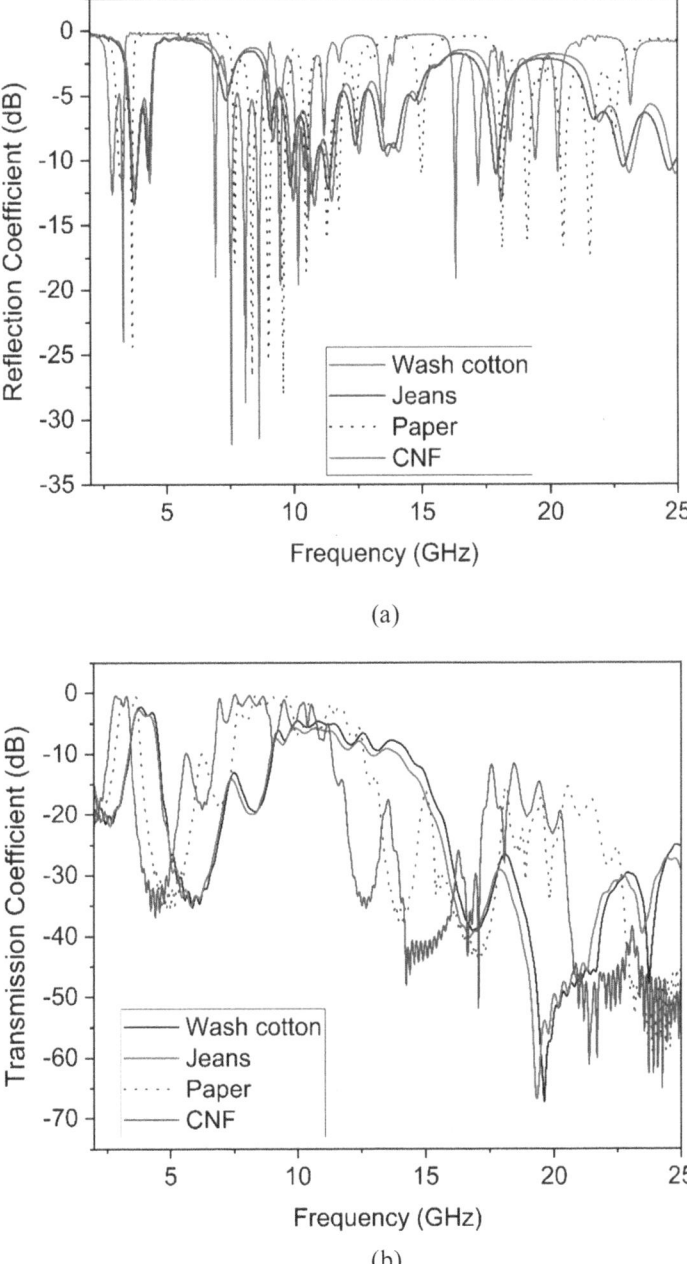

Fig. 6. S-parameters with different flexible substrates (a) Reflection results, (b) Transmission results

scale of 15 A/m, and there is minimum current coupling observed across other section of the sensor.

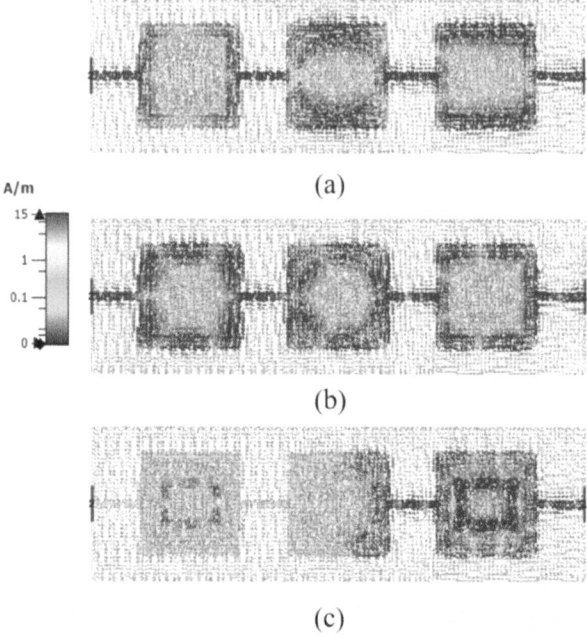

Fig. 7. Surface current density with excited port-1 at the resonated frequency (a) 3.3 GHz, (b) 8.5 GHz, (c) 16.5 GHz

3 The Breast Models Analysis Using HSRR

To carry out the breast model analysis for tumor detection using proposed sensor, there are two models (with tumor and without tumor) of the breast is mimicked as similar property of the actual breast, the breast model parameters are listed in Table 2. However, in this study we have considered layered structure of the breast model ordered as: skin, fat, glandular tissue for normal model, and for malignant model, inserted tumor of small size (10 mm) with relative permittivity of 55 in the glandular tissue. The sensor is positioned in the radiative near field relative to the breast models. However, this region is defined by the equation: $0.62\sqrt{(D^3/\lambda)}$, where λ is the center frequency and D is the maximum dimension of the sensor [17].

Table 2. Breast models parameters

Name of layer	Permittivity (ε_r)	Relative Conductivity (σ) (S/m)	Thickness (mm)
Skin	38	4.5	2
Fat	10	0.5	6
Glandular Tissues	16	0.6	50

We monitored the sensor's simulated S-parameters (reflection/transmission) during interaction with normal/malignant breast models, including free space results. The findings of this analysis are illustrated in Fig. 8. The minimum values deviation and difference between these values for normal and malignant phantom are collected in Table 3, these results are clear evidence of the identification of malignancy in the breast models.

Table 3. S-parameters deviation for normal and malignant breast models

	Normal model	Malignant model	Frequency shift (Δ f), (GHz)	Deviation in the S-parameters value for normal and malignant models (Δ S)min
$(S11)_{min}$ (dB)	-37.92	-29.17	-	8.75
Frequency point (GHz)	9.66	14.85	5.19	-
$(S21)_{min}$ (dB)	-53.75	-54.1	-	0.35
Frequency point (GHz)	24.3	22.7	1.6	-

Among the reflection results, the minimum values for the normal and malignant models were observed as -37.92 dB and -29.17 dB, at frequencies of 9.66 GHz and 14.85 GHz, respectively. There is 8.75 dB difference is observed in the reflection result with the frequency shift 5.19 GHz. However, transmission coefficient (S21) has shown 0.35 dB difference in the (S21) magnitude values with frequency shift 1.6 GHz.

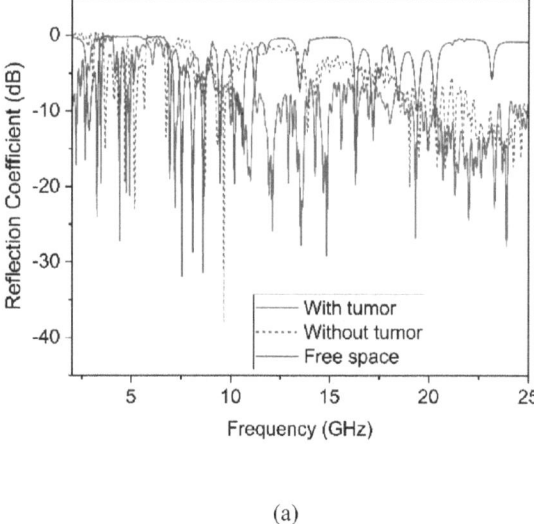

(a)

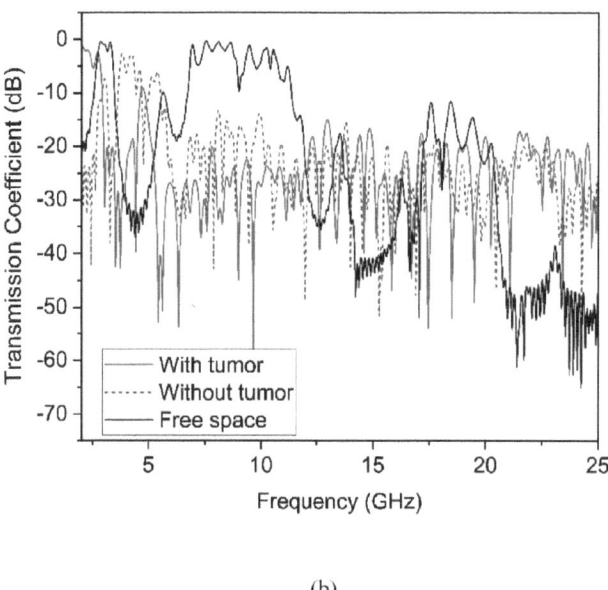

(b)

Fig. 8. S-parameters results for breast models analysis

4 Conclusion

This work has been implemented a multi-resonating flexible and biodegradable sensor for small size tumor assessment. The sensor is implemented on the low-cost plant-based CNF based substrate. The resonating frequency accomplished by the sensor are 3.2 GHz, 6.9 GHz, 7.5 GHz, 8.09 GHz, 9.4 GHz 10.18 GHz, 16.3 GHz, 17.5 GHz, 19.4 GHz, and 20.3 GHz. The surface current distribution and transmission coefficients of the sensor are also verified which is in the acceptable range. The normal and malignant breast models are exposes by the sensor which shown significant difference between the S-parameter results for each case. The proposed sensor has the potential to become a new way to detect breast tumors. Current limitation of the proposed work is, sensor is stationary around the breast model, in future, prototype of the sensor can be fabricated and experiment with realistic breast phantoms and results will assess with either rotating breast model 360 degree or sensor will be move around the object.

Acknowledgments. We would like thank to research unit head and all administration and academic staff to accomplish the research.

Funding. This research is carried out by faculty of College of Engineering at the University of Buraimi, Oman, under the Internal Research Grant (IRG/UoB/CoE-002/2022–23).

Disclosure of Interests. The design and execution of this research are free from any conflicts or competing interests.

References

1. Heisey, R., et al.: Health care strategies to promote earlier presentation of symptomatic breast cancer: current oncology 2011. **18**(5), 227–237 (2011). https://doi.org/10.3747/CO.V18I5.869
2. Heller, S.L., et al.: Breast MRI screening: benefits and limitations. Curr. Breast Cancer Rep **8**(4), 248–257 (2016). https://doi.org/10.1007/S12609-016-0230-7
3. Jaglan, P., et al.: Breast cancer detection techniques: issues and challenges. J. Inst. Eng. (India) **100**(4), 379–386 (2019). https://doi.org/10.1007/S40031-019-00391-2
4. Chandra, R., et al.: On the opportunities and challenges in microwave medical sensing and imaging. IEEE Trans. Biomed. Eng. **62**(7), 1667–1682 (2015). https://doi.org/10.1109/TBME.2015.2432137
5. Charles, J.: Advancing microwave-based imaging techniques for medical applications in the wake of the 5g revolution .In: IEEE Xplore, 2019 13th European Conference on Antennas and Propagation (EuCAP)
6. Islam, M.T., et al.: A low cost and portable microwave imaging system for breast tumor detection using ultra-wideband directional antenna array. Sci. Rep. **9**(1), 1–13 (2019). https://doi.org/10.1038/s41598-019-51620-z
7. Aldhaeebi, M.A., et al.: Dipole array sensor for microwave breast cancer detection. IEEE Access **11**, 91375–91384 (2023). https://doi.org/10.1109/ACCESS.2023.3304694

8. Zhang, H.: Microwave imaging for breast cancer detection: the discrimination of breast lesion morphology. IEEE Access **8**, 107103–107111 (2020). https://doi.org/10.1109/ACCESS.2020.3001039
9. Das, C., et al.: A novel miniaturized millimeter wave antenna sensor for breast tumor detection and 5G Communication. IEEE Access **10**, 114856–114868 (2022). https://doi.org/10.1109/ACCESS.2022.3216858
10. Kaurav, P., et al.: Sub-terahertz waveguide iris probe for ex-vivo breast cancer tumor margin assessment. IEEE J. Electromagnet. RF Microw. Med. Biol. **6**(3), 406–412 (2022). https://doi.org/10.1109/JERM.2022.3172859
11. Elsheakh, D.M., et al.: Textile monopole sensors for breast cancer detection. Telecommun. Syst. **82**(3), 363–379 (2023). https://doi.org/10.1007/S11235-023-00990-X/TABLES
12. Alibakhshikenari, M., et al.: Metamaterial-inspired antenna array for application in microwave breast imaging systems for tumor detection. IEEE Access **8**, 174667–174678 (2020). https://doi.org/10.1109/ACCESS.2020.3025672
13. Kumar Sharma, M., et al.: Noninvasive microwave-multielement sensor for breast phantoms analysis and tumor detection. IEEE Sens. J. **23**(17), 20207–20214 (2023). https://doi.org/10.1109/JSEN.2023.3296740
14. Salvador, et al.: Experimental tests of microwave breast cancer detection on phantoms. IEEE Trans. Antennas Propag. **57**(6), 1705–1712 (2009)
15. Liu, et al.: Tissue phantom-based breast cancer detection using continuous near-infrared sensor. Bioengineered **7**(5), 321–326 (2016)
16. Khajeh-Khalili, F., et al.: A novel wearable wideband antenna for application in wireless medical communication systems with jeans substrate. J. Text. Inst. **112**(8), 1266–1272 (2021). https://doi.org/10.1080/00405000.2020.1809909
17. Yaghjian, A.D.: An overview of near-field antenna measurements. IEEE Trans. Antennas Propag. **34**(1), 30–45 (1986). https://doi.org/10.1109/TAP.1986.1143727

XR-Menu: Food Ordering System in Extended Reality Application

Ajune Wanis Ismail[✉] and Fazliaty Edora Fadzli

Mixed and Virtual Reality Research Lab, ViCubeLab, Faculty of Computing, Universiti Teknologi Malaysia, 81310 Johor, Malaysia
{ajune,fazliaty.edora}@utm.my

Abstract. Food ordering system in extended reality (XR) technology refers to a comprehensive system or application that utilizes XR technologies such as augmented reality (AR), virtual reality (VR), or mixed reality (MR) to enhance the food ordering process. This system may incorporate various XR elements, such as interactive menus, virtual dining experiences, or immersive environments, to allow users to browse menu items, customize orders, and visualize dishes before placing an order. It typically involves the integration of XR technology into the entire food ordering workflow, from menu selection to payment and order fulfillment. However, the problem with VR implementation is scene complexity in VR that creates motion sickness and interaction in AR only will limits the user preferences. This paper explains the development state to develop a food ordering system in XR application. This study presents the preliminary study on the previous works for AR in the food ordering system. Then, it explains the design of the XR framework for the food ordering system. Finally, this research shows the integration results of the XR framework into the food ordering system. XR menu specifically refers to a digital menu that leverages XR technologies to enhance the browsing and selection of menu items. Unlike a full-fledged food ordering system, an XR menu may focus solely on presenting menu options in an immersive or interactive way using AR, VR, or MR. XR technology has enhanced the user experience and using XR to improve the conventional food ordering system.

Keywords: Human Computer Interaction · Augmented Reality · Virtual Reality

1 Introduction

A food ordering system facilitates swift and user-friendly order placement. However, it's worth noting that some restaurants lack attractive User Interface (UI) and a satisfying User Experience (UX) [1]. Many smaller restaurants may not have such systems in place and may not offer appealing menus on their websites or through third-party food ordering platforms. Online food ordering systems leverage technologies such as Augmented Reality (AR), Virtual Reality (VR), and Extended Reality (XR) to enhance customer engagement. XR combines the capabilities of VR and AR to create immersive experiences with multidimensional information, elevating user engagement in emerging consumer industries [2]. Both concepts involve the use of XR technologies in the context of food service, a food ordering system using XR technology encompasses a broader range of functionalities and features, including menu browsing, customization, ordering, and payment, whereas an XR menu specifically refers to a digital menu enhanced with XR elements for a more immersive browsing experience.

AR technology links computer-generated elements with the real world, enabling real-time interactions [3]. However, precise tracking is essential for the accurate alignment of digital content. The most significant static registration errors stem from inaccuracies in tracking and sensor device outputs [4]. Currently, AR is predominantly displayed on handheld devices like mobile phones or smartphones, but imprecise location and orientation sensors can decrease accuracy and lead to orientation data drift [5].

VR technology offers an immersive experience, enabling users to interact with a computer-generated three-dimensional (3D) world and simulating one or more of their five senses in real-time [6]. Most users require a head-mounted display (HMD) to fully engage with VR. However, some users may experience side effects such as nausea, eyestrain, visual disturbances, loss of balance, headaches, and drowsiness, which can potentially make them uncomfortable and deter future VR use [7]. This paper starts with problem background, followed by a detailed explanation of the proposed method. Subsequently, the results of the study are presented, leading to a discussion of the findings in the conclusion. The paper ends with a conclusion and implications drawn from the study's outcomes.

2 Problem Background

Online food ordering is now facilitated by AR, a technology that can display menus, cuisine options, and personalization features within restaurants [8]. This means that customers can thoroughly browse menus, explore food items, and estimate portion sizes before placing an order. When customers can visualize a food product in AR, it enhances their satisfaction as the item closely matches their expectations. Additionally, as observed by Batat [8], AR can significantly boost immersion in the restaurant experience, elevating client enjoyment, curiosity, hedonism (experience-based enjoyment), and overall well-being during the meal.

On the other hand, VR provides an even more lifelike experience, surpassing static images in evoking emotional responses in humans [9]. This suggests that virtual representations of food, as opposed to photographs of real food, can effectively enhance customer satisfaction with specific meals. By implementing VR technology, retailers can enrich the shopping experience and foster customer loyalty [10]. The same applies to food ordering, where the customer's ordering experience can be significantly enhanced.

An example of AR's application in food ordering is KabaQ, an innovative application that elevates restaurant menus. KabaQ empowers customers with informed choices, meeting their expectations and enabling them to view food from multiple perspectives. It also allows customers to create personalized meals, as demonstrated in Fig. 1(a) with the usage of the KabaQ application.

JOMMAKAN! a mobile food ordering system built for clients in a restaurant to order food, is another earlier work by [12]. The system allows you to display menus, order,

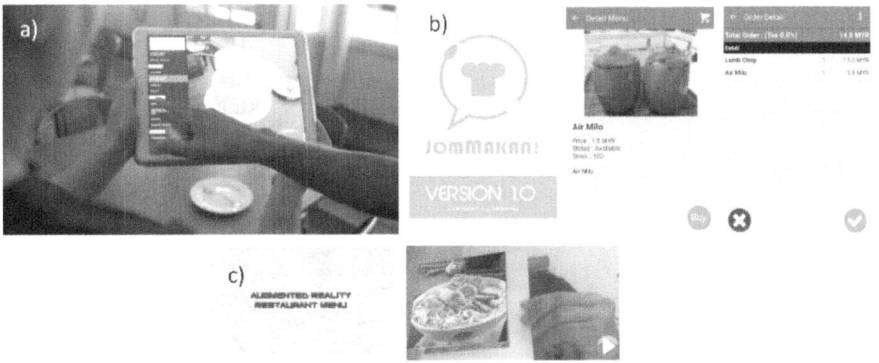

Fig. 1. Previous work of (a) Kabaq application [11, 12]; (b) JOMMAKAN! [12]; (c) AR Restaurant Menu [13]

make purchases, checkup profiles and information, share your system, and contact the administrator. The menu display and food ordering are the main focus of the consumer interface. The customer can select and order their desired food. Figure 1(b) depicts the JOMMAKAN! loading interface, interface for the selected food, and interface for the ordered list.

Another previous by [13] is titled "Augmented Reality Restaurant Menu." The utilization of AR technology enables the presentation of menus in a very detailed and engaging manner, incorporating elements such as animation and video. Figure 1(c) depicts the loading interface of the work, the trackable design of the work and video control design of the work.

The comparison between the previous research studies and the current study is presented in Table 1. In Table 1, the research incorporates extended reality technology of AR and VR, whereas the remaining applications solely employ a single technology of AR. The research findings presented in this study have implications for the development and implementation of web-based applications. According to [14] Web-based applications offer a convenient means of access for individuals, eliminating the need for application installation on mobile devices. This research also enables interactivity in the application for users to interact.

Table 1. Comparison between KabaQ, JOMMAKAN!, AR Restaurant Menu and this reseach.

	KabaQ	JOMMAKAN!	AR Restaurant Menu	This Research
Technology Used	AR	No technology	AR	AR/VR
Type of Application	Mobile-based	Mobile-based	Mobile-based	Web-based
Interactivity	Interactive	Not interactive	Interactive	Interactive

Ground and plane detection tracking are frequently employed by AR developers through the utilization of ARCore and ARKit Software Development Kits (SDKs). ARCore is an open-source and freely available SDK that offers compatibility across a wide range of mobile devices, hence offering a distinct edge in facilitating AR experiences. The comparative study conducted by [15] provides a comparison between ARCore and ARKit, as presented in Table 2.

Table 2. Comparison between ARCore and ARKit

	ARCore	ARKit
License	Open source and free	Not free and open to all
Platform	Android	iOS
Ground & plane detection	Stable	Stable

Four types of HMDs are used to visualize VR experiences. Oculus Rift, HTC Vive, and Samsung HMD Odyssey offer better immersion than Google Cardboard. The former three devices have moderate to difficult mobility and setup, while Google Cardboard is easy. Extended reality (XR) requires a mobile phone configuration, and Google Cardboard is better for VR. Table 3 shows [16]'s Google Cardboard, Oculus Rift, HTC Vive, and Samsung HMD Odyssey comparison.

Table 3. Comparison between Google Cardboard, Oculus Rift, HTC Vive and Samsung HMD Odyssey

	Google Cardboard	Oculus Rift	HTC Vive	Samsung HMD Odyssey
Type	With mobile phone	With PC	With PC	With PC
Platform	Android, iOS	Oculus Home	SteamVR, VivePort	Windows Mixed Reality
Sense of immersion	Medium	Medium-High	Medium-High	Medium-High
Controller	Magnet	Oculus Touch, Xbox One	Vive controller, PC compatible gamepad	Samsung HMD Odyssey
Portability and setup	Easy	Medium	Hard	Medium

One of the characteristics of the web is its open and standards-based nature as the content can be published and accessed by anyone [17]. Web-based XR, consisting of web-based AR and VR, requires a supported XR device application to expose the underlying AR and VR platforms [17].

Web-based AR enables smartphone users to learn more about AR technology in the simplest possible way without having to do the installation via the webpage [14]. As stated by [14] as well, web-based AR allows retailers to make interactive advertisements that can bring advantages for marketing, sales, and brand exposure. AR.js and Google model viewer are two examples of libraries that can be used for web-based AR. The Google model viewer is simple to use and uses marker-less tracking. In comparison to AR.js, the Google Model Viewer library is hence more appropriate for use. Google model viewer also provides the interaction which gives more advantage. Table 4 presents the comparison between AR.js and Google Model Viewer.

Table 4. Comparison between AR.js and Google Model Viewer

	AR.js	Google Model Viewer
Tracking	Marker-based	Marker-less
Implementation	Requires A-Frame	Easy
Interaction	No	Yes
Compatibility	Android, iOS	Android, iOS

Web-based VR allows users to enjoy and create VR content with panorama pictures or 360 videos using smartphones. [18]. A survey by [18] analyzed students' experiences with different web-based VR libraries, including A-Frame, Three.js, A-Frame only, and Unity. [18] concluded that A-Frame is a simple and easy VR setup compared to Three.js and Unity, offering advantages over Three.js and Unity. In addition, A-Frame supports VR headset as Google Cardboard. As claimed by [19], A-Frame is a cross-platform that supports various VR headsets including the Google Cardboard.

3 Proposed Method

This section provides an in-depth system overview, focusing on two primary components: the XR framework and the food ordering system. Within the XR framework, our design and implementation are segregated into two distinct realms: AR and VR. This encompasses AR tracking, VR tools, AR and VR user interfaces, as well as environments ranging from augmented to fully virtual. Furthermore, the section delves into a comprehensive discussion of the food ordering system, shedding light on its key aspects and functionalities.

3.1 AR Tracking Using ARCore

The employed methodology for tracking is the plane tracking technique. The ARCore application programming interface (API) utilized in this context is derived from the Google Model Viewer. It enables the AR camera to initiate and monitor the plane, hence facilitating the visualization of 3D objects. Figure 2 depicts the marker-less AR tracking technique.

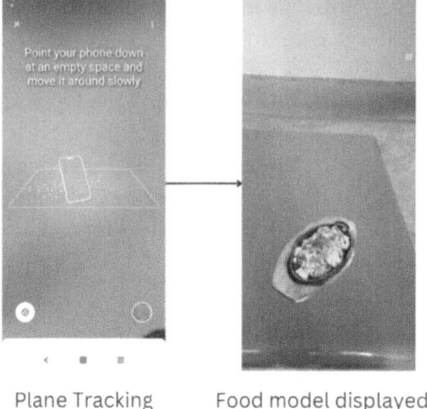

Plane Tracking   Food model displayed

Fig. 2. Marker-less AR tracking

3.2 VR Tools Using Google Cardboard

Google Cardboard is a VR gadget developed by Google that facilitates immersive experiences. In the event that the smartphone possesses the necessary functionalities, Google Cardboard can be utilized as a means to experience VR on said smartphone. In order to operate VR content, Google Cardboard necessitates the presence of a groscope. The gyroscope sensor embedded in smartphones enables users to engage in movement while experiencing VR content. The virtual environment and 3D content were rendered using A-Frame. The visual representation depicted in Fig. 3 illustrates the use of Google Cardboard to operate VR content.

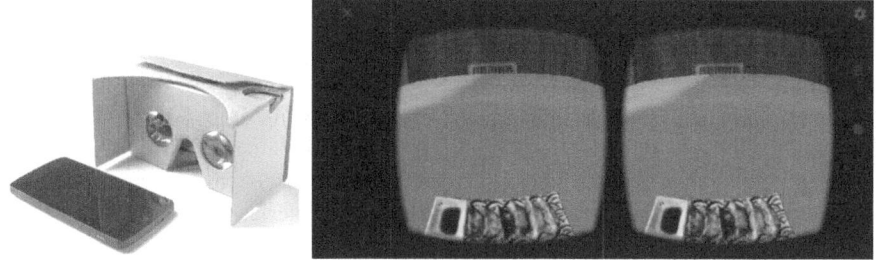

Fig. 3. VR content running using Google Cardboard

3.3 Designing AR User Interface

There are three key components in the UI design of AR, including the AR display, AR tracking, and virtual content. The AR display necessitates the utilization of a camera, which can be acquired from a portable device, provided that it possesses adequate lighting conditions, and the camera is positioned correctly. In this particular instance, a smartphone was employed as the chosen device. The process of AR tracking was conducted utilizing ARCore technology in order to facilitate plane tracking for the purpose of rendering a 3D representation. The virtual content for augmented reality interface design is represented by the food object. The design of the AR interface is depicted in Fig. 4.

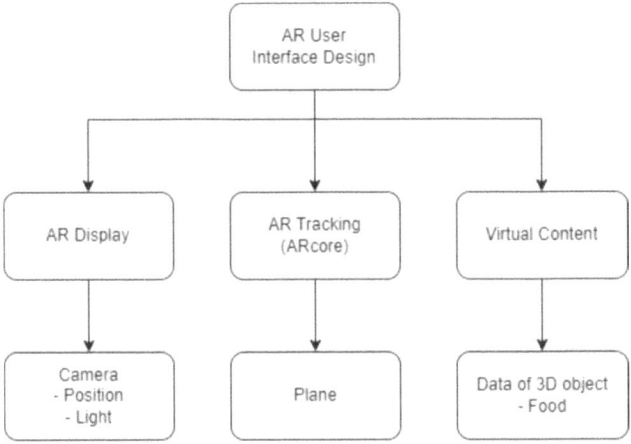

Fig. 4. AR interface design

The AR environment features an augmented 3D food model on SketchFab website, integrated into handheld device with camera for real-world display. Figure 5 illustrates the AR environment for this project.

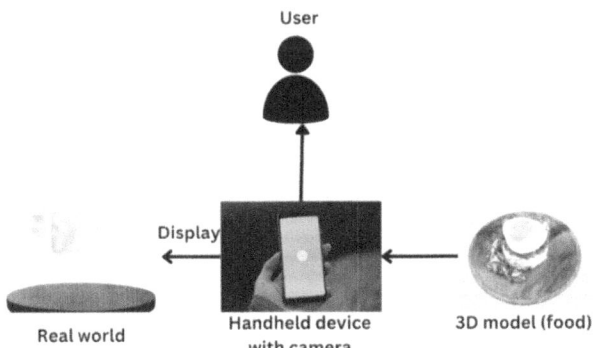

Fig. 5. AR environment

3.4 Designing VR User Interface

There are three primary components that constitute the VR experience: the VR display, VR tools, and virtual content. The utilization of a handheld device, specifically a smartphone, is necessary for the operation of the VR display. VR encompasses the necessary technological components for seeing the VR environment, such as the Google Cardboard platform. The virtual elements utilized in the design of VR interface encompass many components such as food, dining tables, and the surrounding surroundings. Figure 6 depicts the design of the VR interface which consist of VR display, VR tools and content.

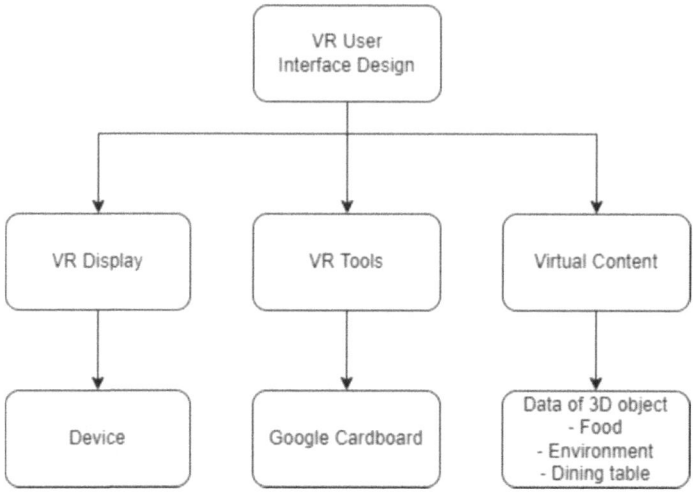

Fig. 6. VR interface design

The virtual environment utilized 3D models sourced from SketchFab. The virtual setting encompasses a 3D model of a restaurant, complete with dining tables. The food model serves as the primary focus within this virtual environment, much like AR The

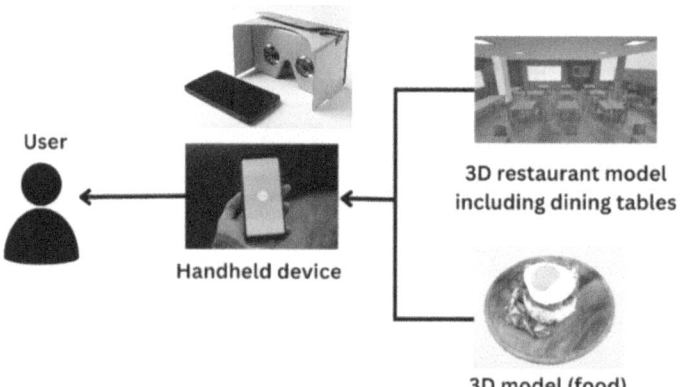

Fig. 7. Virtual environment

integration of food and restaurant models onto a mobile device as a virtual environment was accomplished. Figure 7 depicts the virtual environment utilized in this project.

4 XRMenu Ordering System

The food ordering system offers users a range of functions. These functions empower the user to navigate and utilize the system effectively. Customers initiate the ordering process by visiting the main page. From there, they can explore the list of available restaurants, access their respective menus, and conveniently add items to their cart. The cart displays the selected food items and allows users to make any necessary adjustments before finalizing their order. Once the order is placed, customers can review their order details. A comprehensive list of these customer functions, along with explanations for each, is provided in Table 5.

Table 5. Functions of the customer

Function	Explanation
View the main page	The main page consists of a link to the restaurant list
View restaurant list	View the restaurants that have registered to the system
View menu	View the menu from all the restaurants that were registered
Add cart	Add food to the cart
Edit cart	Edit the cart by changing the quantity of the food
Delete cart	Delete the food in the cart if the quantity becomes zero
View cart	View the food added in the cart
Add order	Add order by placing the order
View order	View the order after sending it to the restaurant

Furthermore, the customer's food ordering system process flow is illustrated in Fig. 8. As shown in the diagram, users initiate the application by scanning the QR code provided on their table. Following a successful scan, users are prompted to enter their table number before gaining access to the list of available restaurants. Once the table number is entered, users can browse the list of restaurants and peruse their respective menus. While viewing the menu, users are given the option to experience it in XR with the choice to select either AR or VR viewing modes. After making their selection, users can input the desired quantity and add the selected items to their cart by clicking the 'Add to Cart' button. In the cart, users can edit their order or proceed to submit it by clicking the 'Place Order' button. Upon submission, users can review the order details they have entered.

This section also demonstrates the results of the web-based food ordering system integrated within the XR application. The XR application leverages web-based technology, specifically utilizing A-Frame, to seamlessly combine Google Model Viewer for AR and A-Frame for VR. This integration offers users a smooth and immersive food

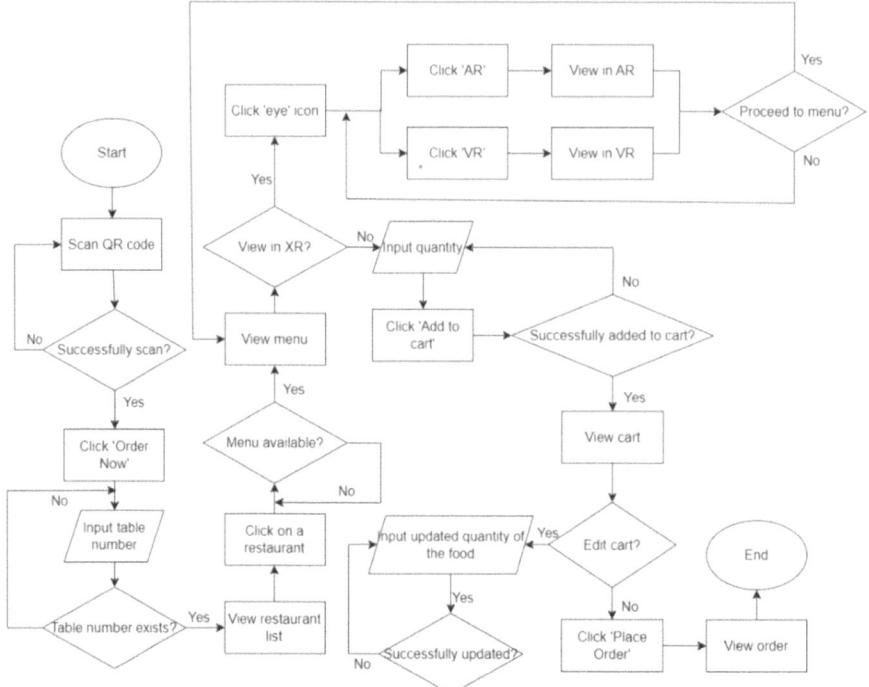

Fig. 8. Customer flow for XR menu application system

Fig. 9. User using the food-ordering system at the food court.

ordering experience. Figure 9 provides a visual representation of a user's experience with the application at the food court.

To use the system, Fig. 10 provides a visual representation of the ordering process, illustrating the step-by-step sequence. It commences with the application's main page and guides users to enter their table number, after which they can initiate the ordering process by selecting 'Order Now.' "After entering their table number, users simply click the 'Let's Go' button, triggering the display of a list of available restaurants.

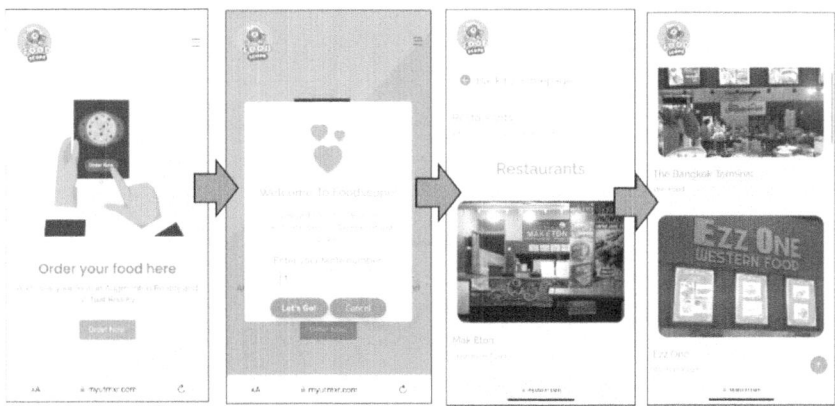

Fig. 10. Ordering Process Flow

In the subsequent phase of the ordering process, users have the flexibility to select the restaurant of their choice, prompting the display of the restaurant's menu. The menu provides detailed information, including the name, description, ingredients, calorie content, and price of each item. To view the menu in AR users simply need to click on the 'eye' icon and opt for the AR viewing mode. Once the 'eye' button is activated, a 3D representation of the menu materializes in AR, offering a visually engaging and interactive menu exploration. The process is visually exemplified in Fig. 11. In Fig. 11, user choose 'chicken tomyam' menu, and the 3D model of the menu appears in AR.

Fig. 11. Viewing menu in AR

In the case of VR viewing, users can easily activate the VR viewing option by clicking on the 'VR View' button. This instantly transforms their selected menu items into an immersive VR environment, offering an engaging and interactive way to explore the menu. The viewing process of VR is depicted in Fig. 12. In Fig. 12, the 3D model of the fried mee menu appears in VR.

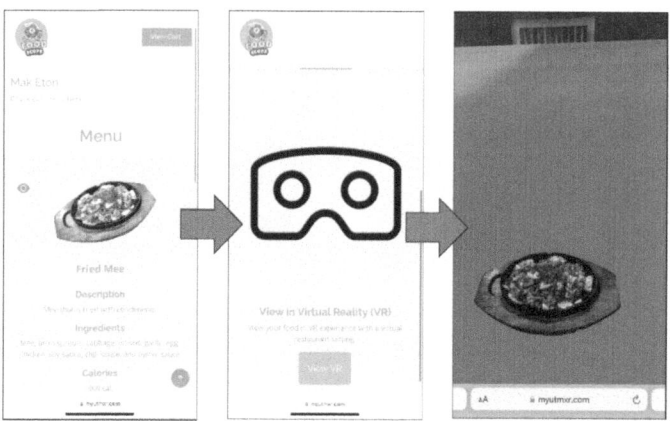

Fig. 12. Viewing menu in VR.

To place an order for the selected food items, users can easily enter the desired quantity in the provided input field and then click on the 'Add to Cart' button. Upon doing so, a prompt box will appear, confirming the successful placement of the order. To proceed with the order submission, users can click on the 'View Cart' button, which will direct them to a checkout page displaying the selected food items. Upon reaching the checkout page, users can click on the 'Place Order' button, which initiates the transmission of the order to the kitchen. Users will receive an order ID number and can wait for the delivery

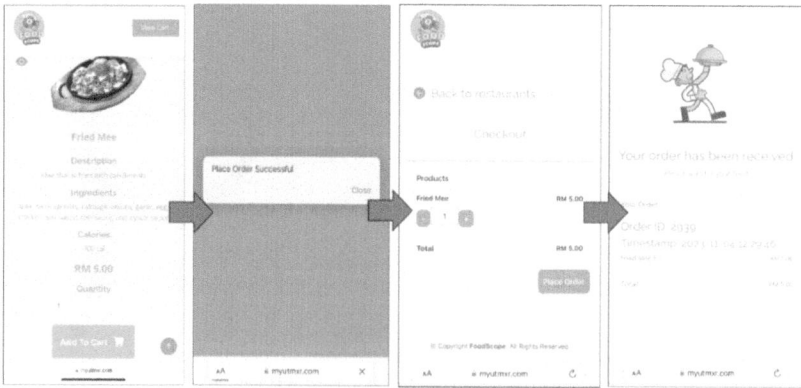

Fig. 13. Order Placement Process Flow

of their chosen dishes to their table. The UI illustrating this process flow can be found in Fig. 13.

5 Conclusion

In this study, we've integrated XR technology into our food ordering system, enhancing user experience, and providing valuable insights for XR application development. This research is significant for the AR/VR field, as it emphasizes practical guidelines for creating XR applications and demonstrates the online, web-based nature of our system. We have shown that well-lit environments and proper phone positioning are critical for efficient tracking. Our work opens up exciting possibilities for the food industry by bridging the digital and physical dining experience through XR technology. For an XR menu and a food ordering system using XR technology, several challenges may arise for future work consideration. The adoption of XR technology relies on users having access to compatible hardware, such as AR glasses, VR headsets, or mobile devices capable of running XR applications. Ensuring widespread availability and affordability of these devices can be a challenge. Integrating XR technology with existing food ordering systems, POS systems, and backend infrastructure presents technical challenges. Ensuring seamless communication and data synchronization between XR interfaces and backend systems is essential for smooth operation. XR applications require efficient use of hardware resources to deliver smooth and immersive experiences. Optimizing rendering performance, minimizing latency, and managing resource consumption are ongoing challenges, particularly for mobile XR devices with limited processing power. Delivering high-quality 3D models, textures, and animations over the web requires efficient content delivery mechanisms. Optimizing asset compression, implementing streaming techniques, and leveraging content delivery can help minimize loading times and bandwidth usage. In conclusion, the development and implementation of XR menus and web-based XR technologies in the food ordering industry offer promising opportunities to enhance the dining experience for customers. Through the integration of XR technologies, users can enjoy more interactive, immersive, and personalized experiences when browsing menus, customizing orders, and visualizing dishes.

Acknowledgement. We would like to extend our heartfelt gratitude to Ministry of Higher Education for the financial support under Fundamental Research Grant Scheme FRGS/1/2023/ICT10/UTM/02/2. Thanks to ViCubeLab at Universiti Teknologi Malaysia for their invaluable support and facilities.

References

1. Brewer, P., Sebby, A.G.: The effect of online restaurant menus on consumers' purchase intentions during the COVID-19 pandemic. Int. J. Hosp. Manag. **94**, 102777 (2021)
2. Abd Karim Ishigaki, S., Ismail, A.W.: Multi-scale avatars in a shared extended reality between AR and VR users. In: International Conference on Expert Clouds and Applications, pp. 573–588. Singapore: Springer Nature Singapore (2022). https://doi.org/10.1007/978-981-99-1745-7_42

3. Rese, A., Baier, D., Geyer-Schulz, A., Schreiber, S.: How augmented reality apps are accepted by consumers: a comparative analysis using scales and opinions. Technol. Forecast. Soc. Chang. **124**, 306–319 (2017)
4. Azuma, R.T.: A survey of augmented reality. Presence: Teleoperators Virtual Environ. **6**(4), 355–385 (1997)
5. Kruijff, E., Swan, J.E., Feiner, S.: Perceptual issues in augmented reality revisited. In: 2010 IEEE International Symposium on Mixed and Augmented Reality, pp. 3–12. IEEE (2010).
6. Guttentag, D.A.: Virtual reality: applications and implications for tourism. Tour. Manage. **31**(5), 637–651 (2010)
7. Weiss, P.L., Jessel, A.S.: Virtual reality applications to work. Work **11**(3), 277–293 (1998)
8. Batat, W.: How augmented reality (AR) is transforming the restaurant sector: investigating the impact of "Le Petit Chef" on customers' dining experiences. Technol. Forecast. Soc. Chang. **172**, 121013 (2021)
9. Gorini, A., Griez, E., Petrova, A., Riva, G.: Assessment of the emotional responses produced by exposure to real food, virtual food and photographs of food in patients affected by eating disorders. Ann. Gen. Psychiatry **9**(1), 1–10 (2010)
10. Kim, J.H., Kim, M., Park, M., Yoo, J.: How interactivity and vividness influence consumer virtual reality shopping experience: the mediating role of telepresence. J. Res. Interact. Mark. **15**(3), 502–525 (2021)
11. Ercan, F.: An examination on the use of immersive reality technologies in the travel and tourism industry. Bus. Manage. Stud. Int. J. **8**(2), 2348–2383 (2020)
12. Azahari, M.H., Ali, F.A.H.: The development of an online food ordering system for JomMakan restaurant. Appl. Inf. Technol. Comput. Sci. **3**(1), 369–376 (2022)
13. Hoe, L.W.: Mobile For Restaurant Menu With Augmented Reality [Unpublished undergraduate thesis]. Universiti Teknologi Malaysia (2015)
14. Nitika, N., Sharma, T.K., Rajvanshi, S., Kishore, K.: A study of augmented reality performance in web browsers (WebAR). In: 2021 2nd International Conference on Computational Methods in Science & Technology (ICCMST), pp. 281–286. IEEE (2021).
15. Hanafi, A., Elaachak, L., Bouhorma, M.: A comparative study of augmented reality SDKs to develop an educational application in chemical field. In: Proceedings of the 2nd International Conference on Networking, Information Systems & Security, pp. 1–8 (2019).
16. Alizadehsalehi, S., Hadavi, A., Huang, J.C.: Virtual reality for design and construction education environment. AEI 2019: Integrated Building Solutions—The National Agenda, 193–203 (2019).
17. MacIntyre, B., Smith, T.F.: Thoughts on the Future of WebXR and the Immersive Web. In: 2018 IEEE International Symposium on Mixed and Augmented Reality Adjunct (ISMAR-Adjunct), pp. 338–342. IEEE (2018)
18. Nguyen, V.T., Hite, R., Dang, T.: Web-based virtual reality development in classroom: From learner's perspectives. In: 2018 IEEE International Conference on Artificial Intelligence And Virtual Reality (AIVR), pp. 11–18. IEEE (2018)
19. Santos, S.G., Cardoso, J.C.: Web-based virtual reality with a-frame. In: 2019 14th Iberian Conference on Information Systems and Technologies (CISTI), pp. 1–2. IEEE (2019).

Predicting Crime Hot Spots Using Machine Learning Algorithms: Cities in USA and South Africa

Dane Brown[(✉)][iD] and Anil Abraham

Rhodes University, Drosty Rd, Grahamstown 6140, South Africa
d.brown@ru.ac.za
http://www.ru.ac.za/computerscience/people/academicstaff/profdanebrown/

Abstract. Crime forecasting is an emerging technology that aids law enforcement in effectively mitigating and responding to crime, using exploratory data analysis and machine learning techniques. No such system exists in South Africa, a country with ever-increasing crime rates. A crime forecasting system developed for Port Elizabeth, a populous city in the Eastern Cape, is proposed to rectify this issue. Additionally, two further forecasting systems were developed using the well-known Boston and Chicago crime datasets. This was done to compare the performance of each system depending on the input dataset. The k-Nearest Neighbour and categorical Gradient Boosting were the best-performing classifiers with accuracies of 88.4% and 71.16%, respectively. These systems achieved comparable results to the literature and showed that accuracies were within 4% of each other regardless of the dataset. However, the South African dataset was found to be a poor candidate for crime forecasting. Using the South African dataset, the best-performing classifier achieved a low accuracy of 15.58%. The proposed system highlights the urgent need for better database record-keeping for crimes in South Africa.

Keywords: Machine Learning · Crime Forecasting · Hotspot prediction · Predictive Policing

1 Introduction

Crime is any illegal act against another person causing harm or damage to property, which is punishable by law [6]. High crime rates lead to financial and emotional stress and result in a decline in tourism. Furthermore, this increases the cost of law enforcement to combat such crimes. As crime rates have skyrocketed in South Africa over the years, South Africa now holds the third-highest crime rate in the world[1].

---

[1] Crime Rate by Country 2023. (n.d.). Retrieved March 2, 2023, from https://worldpopulationreview.com/country-rankings/crime-rate-by-country.

This work was undertaken in the Distributed Multimedia CoE at Rhodes University.

Crime forecasting, also known as predictive policing, takes data from contrasting sources, analyses them, and uses the results to predict and respond more effectively to subsequent crime [8]. This assists law enforcement with their strategies for mitigating crime while helping predict where future crimes are likely to occur. With the South African Police Service (SAPS) armed with this knowledge, South Africa could see a decline in crimes committed as they could respond to crimes quicker than before.

Machine learning is a highly effective approach to crime forecasting, as it can analyse large volumes of diverse data sources and generate accurate predictions [6]. This lends itself well to predicting where a crime is likely to occur. The challenge is to build various machine learning classifiers, evaluate each based on their resulting performance metrics, and subsequently select the best classifier.

Cities such as Boston and Chicago have attempted to reduce crime rates locally by developing crime forecasting systems, such as Chicago's Citizen Law Enforcement Analysis and Reporting (CLEAR) system. This helps law enforcement to use their resources more efficiently. South Africa could benefit from a robust crime forecasting system similar to those developed for Boston and Chicago due to its ability to predict crime hotspots for specific years. Port Elizabeth, a populous city in South Africa, is used to test this hypothesis.

2 Crime Forecasting

Crime forecasting entails predicting crime before it happens, allowing police resources to be utilised more effectively to mitigate crime better [8]. Techniques such as crime mapping, geospatial prediction, exploratory data analysis, and machine learning are used extensively to develop robust forecasting systems. Crime forecasting has been shown to help inform police management decisions and to help discover hidden trends in the underlying data [7]. The following subsections explore previously implemented crime forecasting systems, including systems that utilised the Boston and Chicago crime datasets.

2.1 History of Crime Forecasting Systems

Yu et al. [15] explored machine learning techniques for crime forecasting. Their goal was to accurately predict the time, location, and probability of future residential burglaries within a United States city. Decision Trees (J48), k-NN (k = 1), and Naïve Bayes were explored. The classifiers used a t-month-based features approach, meaning a burglary in one month can be described by events that happened prior to the burglary. They found that Decision Trees when $t = 2$ yielded an accuracy of 88.33%, with k-NN (k = 1) yielding 87.70% when $t = 10$, where t is measured in months.

Iqbal et al. [5] studied classification algorithms for crime prediction, particularly Naïve Bayes and Decision Trees. Their goal was to predict the crime severity of each USA state based on the number of violent crimes per population, using the 1995 FBI Uniform Crime Reports dataset. Features explored were US state, community population, and number of people under the poverty level. Ten-fold cross-validation was applied. Decision Trees yielded an accuracy and F1-score of 83.95% and 82.6%, respectively.

An experiment by Sun et al. [12] aimed to use crime forecasting to predict criminal types based on specific characteristics about an individual. The Dalian Police Bureau's crime dataset was used. The features included were age, sex, profession, marital status, and other related features. The Naïve Bayes, C4.5 algorithm, and k-Nearest Neighbour (k-NN) models were implemented. C4.5 is a Decision Tree algorithm that builds on the information entropy concept [12]. Each model's performance was evaluated using ten-fold cross-validation. The k-NN algorithm scored an accuracy of 66.69%, with the C4.5 and Naïve Bayes algorithms yielding accuracies of 66.16% and 54.88%, respectively.

Wibowo and Oesman [14] aimed to predict crime types using machine learning models in the Sleman Regency in Java, Indonesia. The Sleman Regency's Police Department's dataset was used, and the models implemented were k-NN, Naïve Bayes, and Decision Trees. The features included were day, time, district, crime type, and other related features. Data preprocessing included data imputation and the removal of outliers and redundant data from the dataset. The Naïve Bayes algorithm performed the best with an accuracy of 65.59%. Decision Trees and k-NN yielded an accuracy of 61.57% and 60.3%, respectively.

Jenga et al. [6] evaluated several modern crime prediction techniques that utilise machine learning. The analysis focused on crime forecasting papers utilizing machine learning, addressing eight key research questions. Most studies aimed to develop novel crime prediction models, with others concentrating on spatio-temporal hotspot forecasting and suspect prediction. Commonly used data sources included crime records, Facebook posts, tweets, and census data, while frequently identified features encompassed crime ID, date, crime category, latitude, and longitude.

Governments' crime record datasets, such as those from Boston and Chicago, were commonly utilized. Evaluation metrics such as accuracy, area under the ROC curve, and precision were prevalent. Supervised machine learning models predominated as the most common machine learning category. Authors commonly cited challenges in data collection, notably the limited public availability of government datasets.

2.2 Boston

The Boston Police Department provided crime incident reports from June 2015 until September 2018. Each record lists information regarding a specific crime incident. This dataset lists over 319,000 records with 17 features, such as district, shooting occurrence, reporting area, location, and date.

R et al. [9] implemented a Boston crime forecasting system using machine learning to aid law enforcement officers with mitigating crime. They aimed to predict crime hotspots by determining which Boston districts contributed the most to crime. Models such as Decision Trees, k-NN, Adaptive Boosting (AdaBoost), Random Forests, and gradient-boosted algorithms were explored due to their effectiveness [2]. Similarly, Sharma et al. [11] aimed to predict the severity of crime in Boston whenever shootings were involved. This experiment used Random Forests and Decision Trees with principle component analysis (PCA). Both systems implemented Exploratory Data Analysis (EDA) algorithms to gain insights into the Boston crime dataset. The features of interest in the two experiments included district, crime severity, reporting area, and date, where crime severity and district were used separately as target variables.

Gradient-boosted algorithms yielded the best results when predicting the district where a crime occurred, with an accuracy of 97.2% [9]. AdaBoost yielded the worst results, with an accuracy of 68%. The EDA showed that District B12 was the most significant crime hotspot, with the months of July and August showing a spike in crimes committed. Random Forests with PCA achieved the best results when predicting a crime's severity [11]. The accuracy and F1-score of the best-performing model were 60% and 56%, respectively. Decision Trees without PCA achieved the worst results, with accuracy and F1-score of 51% and 49%, respectively. EDA revealed that the most frequently occurring crime in Boston was motor vehicle accident responses.

2.3 Chicago

The Chicago dataset is derived from the Chicago Police Department's CLEAR system, listing over 7 million reported crime incidents in Chicago from 2001–2023. There are 23 unique features, including arrest information, location, date, crime description, and related.

Tamir et al. [13] aimed to predict the severity of crimes in Chicago based on whether the suspect was arrested. They used k-NN, AdaBoost, Random Forests, and neural networks to predict the crime severity. In contrast, Safat et al. [10] aimed to predict areas in Chicago with high crime density. The models explored included k-NN, Naïve Bayes, Decision Trees, Random Forests, and eXtreme Gradient Boosting (XGBoost).

The feature selection in both experiments included location, crime description, district, latitude, longitude, location description, year, month, and weekday. Data preprocessing included feature encoding and data cleansing, such as the removal of duplicates. Both systems performed EDA on the Chicago dataset to gain insights into the underlying data.

EDA showed that the most frequently occurring crime in Chicago was theft, where 'street' was the most common crime location description. Crime was shown to be gradually decreasing through the years, with peak crime counts in July. Random Forests yielded the best results when predicting the crime severity in Chicago, with an accuracy of 89%. Neural networks performed the best overall but were not considered due to being computationally expensive. The k-NN algorithm performed the worst overall in this experiment, with an accuracy of 85% [13]. Safat et al. [10] found that when predicting high crime density areas in Chicago, XGBoost performed the best with an accuracy of 94%. Decision Trees performed the worst with an accuracy and F1-score of 66% and 75%, respectively.

2.4 South Africa

The South African crime dataset lists crimes from 2010–2022 and is derived from the SAPS dataset[2]. Each crime record lists the number of crimes that occurred within a year in South Africa, according to the police station it was reported to, the province, and the type of crime reported.

South Africa's 2024 crime index is 75.4, the fifth highest in the world[3]. The causes of the crimes include high levels of poverty, inequality, social isolation, unemployment, and the normalizing of violence. Despite this, there exists no public dataset that is comparable to those in USA.

3 Methods and Materials

A crime forecasting system for Boston, Chicago, and Port Elizabeth was developed. The goal of each system is to accurately predict the location of where a crime occurred. The high and low-level design of the system, implementation algorithms used, performance metrics, and datasets are explored in the subsequent sections.

3.1 High-Level Design

The proposed system can be broken down into two stages: An EDA and a machine learning stage. Figure 1 depicts the high-level design of the proposed system.

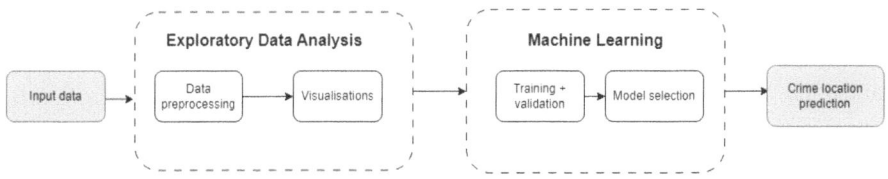

Fig. 1. High level overview of the crime forecasting system

Exploratory Data Analysis. Datasets often contain data with noise, outliers, or missing values. Such data can lead to inaccurate analyses and ineffective machine-learning models. Data preprocessing entails the methods used to transform raw data into an appropriate format, eliminating anomalies such as missing values and outliers [10]. This allows for accurate analyses and robust machine-learning models to be produced. Feature selection, encoding, and data cleaning were utilised.

---

[2] https://www.saps.gov.za/services/crimestats.php.
[3] https://www.statista.com/statistics/1399476/crime-index-south-africa/.

Feature selection involves reducing the number of features in a dataset, which reduces its complexity, allowing for more interpretable analyses and models. Feature encoding involves transforming categorical variables into numerical variables, as most machine learning models can only interpret numerical data. Data cleaning is a form of preprocessing where null values, outliers, and duplicates are removed from a dataset.

Visualisations allow for the discovery of outliers and underlying trends in the data that may not have been apparent. Examples of visualisations include bar graphs, heatmaps, and line charts. Visualisations offer a high-level overview by summarising a dataset and its underlying characteristics.

Machine Learning. Training a machine learning model includes various phases, such as data splitting, hyperparameter tuning, and cross-validation. Data splitting involves splitting the dataset into a train and test set using a predefined ratio. Using a test set allows models to be evaluated, and a data split prevents models from potentially overfitting, which occurs when a model performs well on the training data but poorly on testing data. While several deep learning models can achieve state-of-the-art results, they require feature-rich data with a large number of samples [1]. Hyperparameter tuning is typically necessary in traditional machine learning models to model particular features. This involves selecting and testing different hyperparameter values and monitoring their results [3]. Cross-validation is a technique for evaluating the performance of a machine-learning model on unseen data, which can help to prevent overfitting.

The proposed system is deemed effective if its underlying models accurately predict crime. A comparison must be made between the various models used to evaluate their effectiveness using performance metrics. The best-performing model can then be identified according to its performance across the three datasets.

3.2 Low-Level Design

The low-level design of the system is depicted in Fig. 2. Data often needs to be transformed further for practical use in machine learning models, and such techniques were explored in Sect. 3.1. A data split of 80:20 is used for the training and test set, respectively. Each model is trained using the training data and tested using testing data. Hyperparameter tuning is done during training and ensures models perform optimally. Effective tuning often yields the highest results possible. Finally, each model is evaluated by measuring its performance on the test set.

3.3 Implementation

This section explores the implementation of the crime forecasting system. All code produced was written in Python using libraries such as Pandas, Scikit-learn, and Matplotlib.

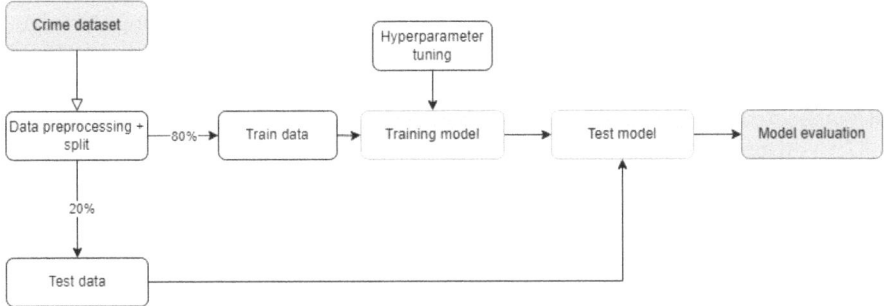

Fig. 2. Low level overview of the crime forecasting system

Datasets and EDA. The forecasting systems each accept a crime dataset as input, where the input is the Boston, Chicago, and South African crime datasets. The Boston and Chicago datasets were detailed in Sect. 2.

Bar charts, line graphs, and similar visualisations were implemented to aid the EDA. Heatmaps of crime hotspots were used to analyse spatial trends in crime. Crime trends and crime types were explored using the above visualisations for the three chosen datasets.

Hyperparameter Tuning and Cross-Validation. The system utilises the following preprocessing methods: Feature encoding, feature selection, and data cleaning. Feature scaling was considered for distance-based algorithms. However, this method was disregarded as each model saw a decrease in accuracy after applying feature scaling. An `OrdinalEncoder` was used for feature encoding, which encodes categorical features into integer arrays. Feature selection is explored in Sect. 4, and the data cleaning methods are detailed in Sect. 3.1.

All models were trained and tested using 5-fold cross-validation and hyperparameter tuning. Five or ten folds are often optimal, but five folds were selected as models such as Categorical Boosting (CatBoost) and XGBoost, explored in Sect. 3.3, are computationally intensive. This means that an increase in folds significantly increases the time and resources spent performing cross-validation. The `RandomizedSearchCV` method was used to implement cross-validation and hyperparameter tuning, using accuracy as the scoring metric. This method samples a random subset of hyperparameters from a given hyperparameter space.

Classifiers. Popular crime prediction models were identified in Sect. 2. As such, the chosen models include Random Forests, k-NN, Decision Trees, Gaussian Naïve Bayes (GNB), CatBoost[4], XGBoost, and AdaBoost. Only classification models were considered, as each system predicts where a crime was committed.

---

[4] The CatBoost model did not require feature encoded data, as it can handle categorical data internally.

Artificial neural networks were also identified as one of the most popular algorithms for crime forecasting in Sect. 2.1. Due to the lack of data for the SAPS dataset and the computationally expensive nature of neural networks, they were not considered further. Where applicable, each model's `random_state` was set to 42 to ensure reproducible results.

3.4 Performance Metrics

The performance of a model is given by its performance metrics. The majority of crime forecasting studies utilised classification models and used accuracy, precision, recall, and F1-score as metrics (Sharma et al. [11], Safat et al. [10]). These metrics are explained in the following subsections.

Accuracy. Accuracy is the most common metric and is a measurement of the ratio between the number of correctly classified predictions and the total number of predictions. The equation for accuracy is shown below. A high accuracy often implies that a model is performing well. However, this is not true for class-imbalanced datasets [4]. A model may be unable to predict classes that do not appear frequently.

$$Accuracy = \frac{\text{Number of correct predictions}}{\text{Total number of predictions}} \quad (1)$$

Precision and Recall. Precision is a measurement of the proportion of identified positives that were actually correct or the accuracy of the positive predictions [4]. The equation for precision is shown below.

$$Precision = \frac{TP}{TP + FP} \quad (2)$$

Recall, also known as sensitivity, is the ratio of positive instances that were correctly detected [4]. The equation for recall is shown below.

$$Recall = \frac{TP}{TP + FN} \quad (3)$$

F1-Score. F1-score is the harmonic mean of the precision and recall of a model. The equation for F1-score is shown below. F1-score is useful for gauging a model's performance, especially when the data of interest has an uneven class distribution. The F1-score often favours classifiers with similar precision and recall values due to the precision/recall trade-off [4].

$$F1 = \frac{2 * Precision * Recall}{Precision + Recall} \quad (4)$$

3.5 Datasets

The following subsections explore the Boston, Chicago, and South African crime datasets.

Boston. This dataset is derived from the Boston Police Department, with records beginning in June 2015 and ending in September 2018. The Boston dataset contains 319,000 crime records with 17 features. Each record listed details a single reported crime. Table 1 depicts the features given in the Boston dataset.

Table 1. Boston dataset features overview

Feature	Description
INCIDENT_NUMBER	File number
OFFENSE_CODE	Code of specific crime
OFFENSE_CODE_GROUP	Name of crime
OFFENSE_DESCRIPTION	Detailed information about crime
DISTRICT	Boston neighbourhood
REPORTING_AREA	Area defined by Boston PD
UCR_PART	Severity of the crime (highest is 1)
SHOOTING	Whether a shooting occurred
DATE/YEAR/MONTH/DAY/HOUR	Date of crime
STREET/Lat/Long/Location	Location

The features of interest are REPORTING_AREA, UCR_PART, OFFENSE_CODE, and DISTRICT. OFFENSE_CODE describes the crime taking place. UCR_PART refers to the severity of a crime, where 'Part One' crimes are the most severe and 'Part Three' crimes are the least severe. The REPORTING_AREA feature is the area defined by Boston Police Department. The DISTRICT feature offers information about the district in which the crime was committed, such as District B3.

Chicago. This dataset is derived from Chicago's CLEAR system with records ranging from 2001–2023. This dataset contains 23 features with over 7 million reported crimes. Each record listed details a single reported crime. Table 2 depicts the features given in the Chicago dataset.

The features of interest are Primary Type, Description, Location Description, Arrest, Domestic, and District. Primary Type and Description describe the type of crimes committed in Chicago. Location Description describes where the incident occurred, such as in a street or apartment. Arrest describes whether the committed crime led to an arrest, and Domestic details whether the crime was domestic-related. The District feature details the various districts within Chicago, with district codes such as 10 and 3.

Table 2. Chicago dataset features overview

Feature	Description
Unnamed: 0	Unknown feature
ID	Unique crime identifier
Case Number	Unique case number recorded by Chicago PD
Date	Estimate of when the incident occurred
IUCR	Illinois Uniform Crime Reporting code
Primary Type	Primary description of IUCR code
Description	Secondary description of IUCR code
Location Description	Description of where the incident occurred
Arrest	Whether an arrest was made
Domestic	Whether incident was domestic-related
Beat	Smallest police geographic area
FBI Code	Classification of crime
Year	Year that incident occurred
Updated On	Date that incident was last updated
Block/Latitude/Longitude/Location	Location
District/Ward/Community Area	Location
X Coordinate/Y Coordinate	Location

South Africa. The official SAPS website releases annual crime statistics for South Africa, with records ranging from 2010–2022. This dataset contains 346,000 records with nine features. There are 5,070 records returned when filtering the dataset for Port Elizabeth. Unlike the Boston and Chicago datasets, the SAPS dataset lists records as yearly crime counts per police station. This means that crimes are not listed at the incident level, nor are the exact locations of reported crimes given. Instead, the police station to which the crime was reported and its respective coordinates are given. Table 3 details the features given in the SAPS dataset.

The features of interest are `Crime`, `Crime Category`, `Crimes`, `Police Station`, and `Year`. `Crime` and `Crime Category` features describe the crime committed. `Police Station` refers to the name of the police station where a crime was reported, such as Humewood. `Crimes` is the total number of reported crimes in a given year per crime type and police station.

Table 3. SAPS dataset features overview

Feature	Description
Crime	Crime description
Crimes	Number of crimes committed per year
Police Station	Name of police station
Province	Name of province
Crime Category	Binning of crime descriptions
Latitude/Longitude/'Latitude, Longitude'	Location
Year	Year that crime occurred

4 Results and Discussion

This section details the experimental setup, presents the results obtained by the EDA and machine learning implemented for the experiments, and concludes with a discussion of these results.

4.1 Experiments

Four experiments were conducted to test the effectiveness of the proposed crime forecasting system using EDA and machine learning. The ability of the system to predict a crime's location for each of the three datasets will be evaluated according to its performance metrics.

Experiment 1. Experiment 1 utilises the Boston crime dataset which lists crime records from 2015–2018. The aim was to predict the district where each crime occurred. As such, the target variable selected was DISTRICT. The Boston dataset's features were detailed in Table 1. This experiment used the following features: REPORTING_AREA, UCR_PART, MONTH, DAY, and HOUR. This feature selection was inspired by the experiment by Sharma et al. and R et al.. A crime and hotspot analysis was conducted as part of the EDA.

Experiment 2. Experiment 2 utilises the Chicago crime dataset, which lists crime records from 2001–2023 and contains over 7 million records. This dataset was sampled to return records from 2012–2017 to improve computational efficiency. This sampling results in around 1,456,714 data records. The goal was to predict the district where each crime occurred, with the target variable set as District. The Chicago dataset's features were outlined in Table 2. This experiment used the following features: Primary Type, Description, Location Description, Arrest, Domestic, and Ward. This feature selection was motivated by the experiment done by Tamir et al.. The EDA was conducted using a crime trend and hotspot analysis.

Experiment 3. Experiment 3 utilised the SAPS dataset. The dataset was first limited to records relating to Port Elizabeth only. This experiment aimed to predict the police station that crimes were reported to for a given year. Thus, the target variable was Police Station. The SAPS dataset's features were explored in Table 3. This experiment used the following features: Crime, Crimes, Crime Category, and Year. The 'Latitude, Longitude' and Province features were dropped as they are considered redundant. The EDA included an overview of crime trends and a hotspot analysis.

Experiments 1 and 2 aim to predict the district feature, with Experiment 1 having 13 unique districts and Experiment 2 having 24 unique districts. Using the SAPS dataset, 13 unique police stations exist. As such, the district feature was selected as the target variable for Experiments 1 and 2 as it can best map onto a Police Station, with roughly the same neighbourhood area with a similar number of unique classes.

Experiment 4. Experiment 4 aimed to determine whether reducing the number of features or data records in an optimal crime dataset, such as the Boston dataset, would lead to significant performance losses for each classifier. As such, the number of features and data records in the Boston dataset was reduced to match that of the SAPS dataset. This was done to compare the trained classifier's results of the two forecasting systems. To test this hypothesis, the classifiers were first trained on the original 319,000 records found in the Boston dataset while reducing the features only to use OFFENSE_CODE_GROUP, OFFENSE_DESCRIPTION, and YEAR. This was done to map the similar features found in Experiment 3. Thereafter, the number of records in the Boston dataset was reduced to 5,070 to match that of the SAPS dataset while using the same features detailed in Experiment 1. For brevity, the Chicago dataset was not utilised for this experiment.

4.2 Results

This section presents the results of the three experiments. Experiments 1, 2, and 3 implemented both EDA and machine learning, with Experiment 4 only implementing machine learning.

Experiment 1. Figure 3 details the annual crime trends in Boston from 2015–2018. The years 2016 and 2017 have significantly higher crime counts than 2015 and 2018. However, one cannot assume fewer crimes were committed during 2015 and 2018, as the Boston dataset recorded data from June 2015 to September 2018. As such, there is missing data for several months in 2015 and 2018, resulting in a lower yearly crime total.

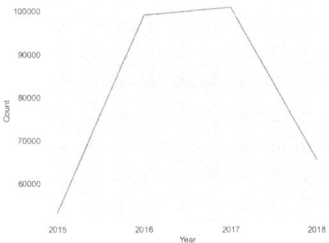

Fig. 3. Line plot depicting annual crime trends in Boston

The most frequently occurring crimes in Boston were motor vehicle accident responses, larceny, medical assistance crimes, investigation, and others, according to Fig. 4. Motor vehicle accident responses are the most frequently occurring crime in Boston. This could be due to the snowfalls in Boston, where driving in the snow becomes significantly more challenging. This leads to an increase in the number of motor vehicle accidents.

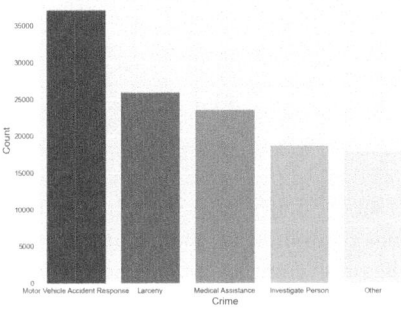

Fig. 4. Bar plot of the most frequently occurring crimes in Boston

A hotspot analysis was conducted to reveal the streets in Boston with the highest number of crimes. A heatmap was created to depict the total crime counts per street, revealing that Washington Street, Blue Hill Avenue, and Boylston Street had the highest incidents of crime in Boston. Figure 5 depicts this heatmap. Many hotspots are converging around downtown Boston. Over 14,000 crimes were committed in Washington Street, and between 4,000–7,000 crimes were committed in the other major hotspots.

The machine learning results of this experiment are shown in Table 4. Ranking refers to ordering the models in terms of their accuracy, from best to worst.

All models performed well predicting a crime's district of occurrence, except for GNB. The best-performing model was k-NN with an accuracy and F1-score of 88.4% and 88.28%. The worst-performing model was GNB, with an accuracy and F1-score of 30.4% and 22.14%. The remaining models performed within

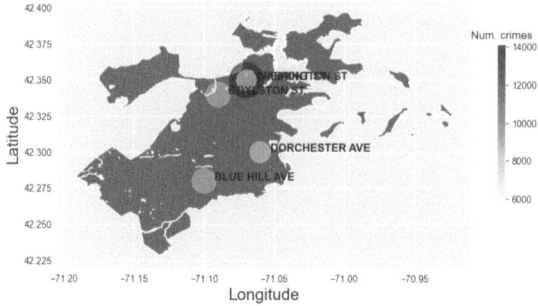

Fig. 5. Heatmap of streets in Boston with the highest number of crimes

Table 4. Evaluation of classification models using Boston dataset

Ranking	Model	Accuracy	Precision	Recall	F1
1	k-NN	**88.4**	88.31	88.4	88.28
2	Random Forests	85.45	85.64	85.45	85.18
3	CatBoost	84.23	84.29	84.23	84.13
4	XGBoost	81.68	81.61	81.68	81
5	Decision Trees	74.54	75.12	74.54	74.66
6	AdaBoost	69.49	68.97	69.49	68.32
7	GNB	**30.4**	**19.81**	30.4	22.14

an accuracy range of 69.49%–85.45%. All but one model, GNB, yielded similar precision and recall values compared to the accuracy achieved. The precision of GNB sees a significant decrease compared to its achieved accuracy and recall values, with a low score of 19.81%. Random Forests are the second best-performing model, which should be noted for later experiments. These results indicate that the Boston crime dataset is a good candidate for use in crime forecasting.

Experiment 2. A rolling sum of all crimes from 2012–2017 was computed to explain whether crime in Chicago decreased over time. Figure 6 shows that crime gradually decreased from 2012–2016 and thereafter started increasing again.

There are 21 unique categories of crimes in Chicago, including criminal trespassing, homicides, and arson. According to Fig. 7, theft, battery, and criminal damage are Chicago's three most common crimes. Theft is the most common crime, suggesting that people are desperate to steal as they may have a family to feed. These findings match those found in the results of Safat et al.'s experiment.

Figure 8 displays the Chicago crime heatmap for use in hotspot analysis, depicting the total crime counts per block. State Street, Michigan Avenue, and Halsted Street are considered major north-south streets due to their significant lengths, and the heatmap revealed that these blocks yielded the highest crime counts in Chicago. Furthermore, the number of crimes committed on these blocks

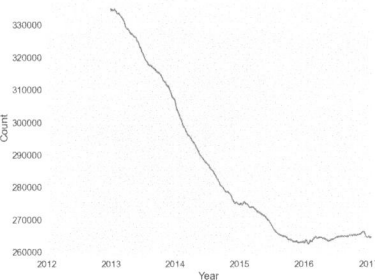

Fig. 6. A rolling sum plot of all crimes in Chicago

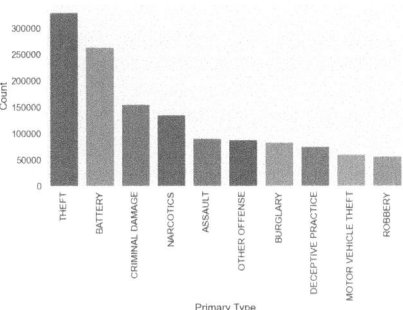

Fig. 7. Bar plot of frequently occurring crimes in Chicago

range from 17,000–23,000. These three blocks account for approximately 8.5% of all crimes in Chicago. These results can aid law enforcement in focusing on these areas through efficient police resource utilisation.

The machine learning results of this experiment are depicted in Table 5.

All models performed well in predicting the target variable except for GNB and AdaBoost. The best-performing model was CatBoost, with an accuracy and F1-score of 71.16% and 70.8%. The worst-performing model was AdaBoost, with an accuracy and F1-score of 23.5% and 17.6%, likely due to overfitting and insufficient weak learners. GNB also performed poorly, with an accuracy and F1-score of 32.81% and 26.23%. The remaining models performed within an accuracy range of 68.38%–71.07%. All models but GNB and AdaBoost yielded similar precision and recall values compared to the accuracy achieved. The precision achieved by GNB and AdaBoost shows a significant decrease compared to its achieved accuracy and recall values, with a low score of 27.67% and 16.72%. The Random Forests classifier appears in the top three best-performing models. These results indicate that the Chicago crime dataset is a good candidate for use in crime forecasting.

Experiment 3. Figure 9 suggests that crime has gradually decreased in Port Elizabeth, particularly in 2020–2022. This decrease in crimes could be due to the

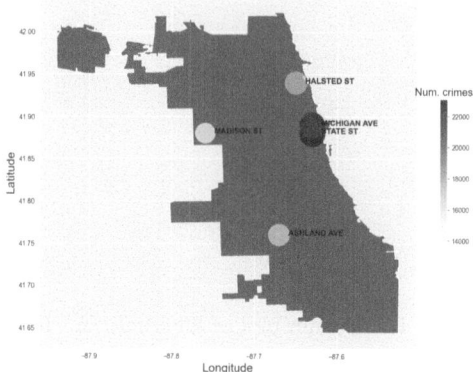

Fig. 8. Heatmap of blocks in Chicago with the highest number of crimes

Table 5. Evaluation of classification models using Chicago dataset

Ranking	Model	Accuracy	Precision	Recall	F1
1	CatBoost	**71.16**	71.46	71.16	70.8
2	XGBoost	71.07	71.45	71.07	70.67
3	Random Forests	70.98	71.23	70.98	70.64
4	Decision Trees	70.47	70.64	70.47	70.19
5	k-NN	68.38	68.32	68.38	68.15
6	GNB	32.81	**27.67**	32.81	26.23
7	AdaBoost	**23.5**	**16.72**	23.5	17.6

Covid-19 pandemic. The years 2010–2014 show a relatively constant crime rate of around 52,500 crimes committed per year. Thereafter, crime gradually decreased from 2015–2020. Crime counts significantly dropped in 2021, with around 36,000 crimes committed, and gradually increased in 2022.

The most frequently occurring crimes in Port Elizabeth from 2010–2022 were theft, residential burglaries, and robberies with aggravating circumstances. This is shown in Fig. 10. These results suggest that theft is the biggest crime that needs the immediate attention of law enforcement.

The hotspot analysis utilised the crime heatmap shown in Fig. 11. The police stations with the most reported crimes in Port Elizabeth were Mount Road, Humewood, and Bethelsdorp. These three police stations account for 34% of all reported crimes in Port Elizabeth, with crime counts ranging between 63,000–82,800 per police station. Law enforcement can use this information to patrol these neighbourhoods more frequently, potentially reducing crime. One cannot identify locations of converging crime hotspots, as each hotspot shown in the heatmap depicts a single police station.

The machine learning results of this experiment are depicted in Table 6.

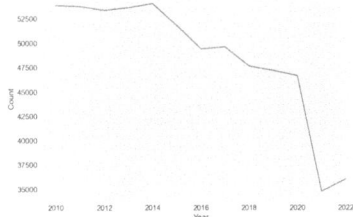

Fig. 9. Line plot depicting annual crime trends in Port Elizabeth

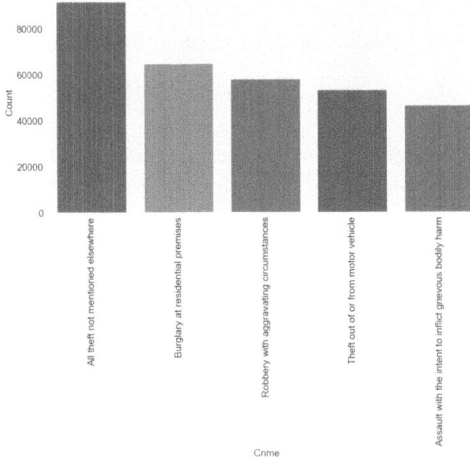

Fig. 10. Bar plot of frequently occurring crimes in Port Elizabeth

All models performed poorly on the SAPS dataset when predicting the police stations where crimes were reported. The best-performing model was XGBoost, with an accuracy and F1-score of 15.58% and 15.32%. The worst model was GNB, yielding an accuracy and F1-score of 8.28% and 4.36%. The remaining models all performed with an accuracy range of 10.75%–15.09%. Precision scores ranged between 5.81%–15.71%, and recall scores ranged between 8.28%–15.58%. Similar to Experiments 1 and 2, the precision scores of AdaBoost and GNB are considerably lower than their corresponding accuracy scores. The Random Forests model again appears in the top three best-performing models. These results indicate that the SAPS dataset is a poor candidate for use in crime forecasting.

Experiment 4. The machine learning results of this experiment are depicted in Table 7. The features column refers to the classifier's results when trained on the Boston dataset, using the features found in Experiment 1 with 5,070 data records. The records column refers to the classifier's results using all 319,000 records and the three features detailed in Sect. 4.1.

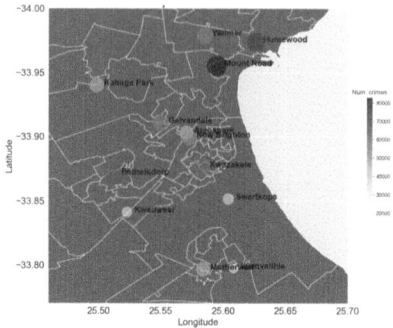

Fig. 11. Heatmap of police stations in Port Elizabeth with the most reported crimes

Table 6. Evaluation of classification models using SAPS dataset

Ranking	Model	Accuracy	Precision	Recall	F1
1	XGBoost	**15.58**	15.71	15.58	15.32
2	Random Forests	15.09	15.22	15.09	15.07
3	k-NN	13.91	14.13	13.91	13.65
4	Decision Trees	13.31	15.13	13.31	13.6
5	CatBoost	12.03	11.91	12.03	11.61
6	AdaBoost	10.75	**6.32**	10.75	6.84
7	GNB	**8.28**	**5.81**	8.28	4.36

The classifiers yield a moderate loss in performance when reducing the number of data records while retaining the same number of features. Experiment 1's best-performing model was k-NN with an accuracy of 88.4%, while Experiment 4's best-performing model was CatBoost with an accuracy of 81.07%. GNB is

Table 7. Evaluation of classification models using Boston dataset when reducing features and data records

Model	Accuracy		Precision		Recall		F1	
	Features	Records	Features	Records	Features	Records	Features	Records
CatBoost	81.07	19.12	81.32	22.06	81.07	19.12	81.0	15.03
XGBoost	76.33	18.38	76.34	12.84	76.33	18.38	75.42	11.68
k-NN	74.06	19.58	73.93	19.48	74.06	19.58	73.38	15.39
Random Forests	66.37	19.63	65.46	19.05	66.37	19.63	63.95	15.4
Decision Trees	46.65	15.93	46.13	4.66	46.65	15.93	46.05	6.24
AdaBoost	41.62	14.39	42.58	13.8	41.62	14.39	37.54	13.34
GNB	30.77	19.68	20.17	18.98	30.77	19.68	23.76	15.06

performing better with fewer data records than Experiment 1's results. There is a significant decrease in performance when using all 319,000 records with fewer features. The classifier's accuracies range between 14.39%–19.68%, indicating that reducing the number of features significantly impacts model performance more than reducing data records.

4.3 Discussion

The classifiers performed the best on the Boston dataset, with the best classifier achieving an accuracy of 88.4%, compared to the best model on the Chicago dataset, achieving an accuracy of 71.16%. The highest F1-score on the Boston dataset was 88.28%, while the highest on the Chicago dataset was 70.8%. This suggests that crime datasets should be recorded similarly to yield high prediction results. These models achieved similar results to those implemented in R et al. and Safat et al.'s experiments. Furthermore, the Random Forests classifier appears in all experiments' top three best-performing models. Thus, Random Forests were identified as the best overall model for crime forecasting.

In contrast, the classifiers performed poorly using the SAPS dataset, with the best classifier achieving an accuracy of 15.58%. The inherently different structure of the SAPS dataset as compared to the Boston and Chicago datasets leads to these poor results. The SAPS dataset has few entries and features, while the Boston and Chicago datasets have over 17 features with abundant entries. Experiment 4 revealed that fewer features in the crime datasets have a more significant impact than fewer data records. This explains the poor results achieved by the classifiers trained on the SAPS dataset, indicating the urgent need for additional complex features, such as the crime's severity and the time the crime was committed.

The EDA revealed that the most frequent crimes in Boston and Chicago occurred between 4 p.m. and 6 p.m., and between 6 p.m. and 8 p.m., respectively. Areas with the most frequent crimes included Washington Street with over 14,000 crimes committed, State Street with over 23,000 crimes committed, and Mount Road with over 82,800 crimes reported. Experiment 3's analysis revealed that 2014 was the peak year for crime in Port Elizabeth. However, it gradually decreased thereafter, and theft was revealed to be the most common crime that needs law enforcement's immediate attention.

5 Conclusion

This study explored various crime forecasting systems developed for Boston, Chicago, and Port Elizabeth. The approach implemented both an EDA and machine-learning phase. The Boston and Chicago crime datasets offered abundant data with complex features. As such, the classifiers performed well using these datasets, with classifiers such as k-NN scoring accuracies up to 88.4% when predicting the district of a committed crime. In contrast, the same models performed poorly on the SAPS dataset. This is due to the inherent differences and simplicity of the SAPS dataset compared to the Boston and Chicago datasets. It was proved that fewer features have a larger impact on the evaluation of a classifier compared to fewer data records. This indicates the need for better data recording for crime in South Africa by adding more complex features resembling those found in the Boston and Chicago datasets.

References

1. Brown, D., Bradshaw, K.: Deep palmprint recognition with alignment and augmentation of limited training samples. SN Comput. Sci. **3**(1), 11 (2022)
2. Brown, D., Sepula, C.: Darknet traffic detection using histogram-based gradient boosting. In: Suma, V., Lorenz, P., Baig, Z. (eds.) Inventive Systems and Control. LNNS, vol. 672, pp. 795–807. Springer, Singapore (2023). https://doi.org/10.1007/978-981-99-1624-5_59
3. Chindove, H., Brown, D.: Adaptive machine learning based network intrusion detection. In: Proceedings of the International Conference on Artificial Intelligence and its Applications, pp. 1–6 (2021)
4. Geron, A.: Hands-on Machine Learning with Scikit-Learn, Keras, and TensorFlow 3E: Concepts, Tools, and Techniques to Build Intelligent Systems, 3rd edn. O'Reilly Media, Sebastopol (2022)
5. Iqbal, R., Murad, M.A.A., Mustapha, A., Panahy, P.H.S., Khanahmadliravi, N.: An experimental study of classification algorithms for crime prediction. Ind. J. Sci. Technol. **6**(3), 4219–4225 (2013). https://doi.org/10.17485/ijst/2013/v6i3.6
6. Jenga, K., Catal, C., Kar, G.: Machine learning in crime prediction. J. Amb. Intell. Hum. Comput. **14**(3), 2887–2913 (2023). https://doi.org/10.1007/s12652-023-04530-y
7. Meijer, A., Wessels, M.: Predictive policing: review of benefits and drawbacks. Int. J. Publ. Adm. **42**(12), 1031–1039 (2019). https://doi.org/10.1080/01900692.2019.1575664
8. Pearsall, B.: Predictive policing: the future of law enforcement. Natl. Inst. Justice J. **266**(1), 16–19 (2010)
9. Anuvarshini, S.R., Nidhi, D., Deeksha Sree, C., Krishna Sowjanya, K.: Crime forecasting: a theoretical approach. In: 2022 IEEE 7th International Conference on Recent Advances and Innovations in Engineering (ICRAIE), vol. 7, pp. 37–41 (2022). https://doi.org/10.1109/ICRAIE56454.2022.10054345
10. Safat, W., Asghar, S., Gillani, S.A.: Empirical analysis for crime prediction and forecasting using machine learning and deep learning techniques. IEEE Access **9**, 70080–70094 (2021). https://doi.org/10.1109/ACCESS.2021.3078117
11. Sharma, H.K., Choudhury, T., Kandwal, A.: Machine learning based analytical approach for geographical analysis and prediction of Boston City crime using geospatial dataset. GeoJournal (2021). https://doi.org/10.1007/s10708-021-10485-4
12. Sun, C., Yao, C., Li, X., Lee, K.: Detecting crime types using classification algorithms. J. Digit. Inf. Manage. **12**(5), 321–327 (2014)
13. Tamir, A., Watson, E., Willett, B., Hasan, Q., Yuan, J.S.: Crime prediction and forecasting using machine learning algorithms. Int. J. Comput. Sci. Inf. Technol. **12**(2), 26–33 (2021)

14. Wibowo, A.H., Oesman, T.I.: The comparative analysis on the accuracy of K-NN, Naïve Bayes, and decision tree algorithms in predicting crimes and criminal actions in Sleman regency. J. Phys. Conf. Ser. **1450**(1), 012076 (2020). https://doi.org/10.1088/1742-6596/1450/1/012076
15. Yu, C.H., Ward, M.W., Morabito, M., Ding, W.: Crime forecasting using data mining techniques. In: 2011 IEEE 11th International Conference on Data Mining Workshops, pp. 779–786. IEEE (2011). https://doi.org/10.1109/ICDMW.2011.56

Experience with the Implementation of Machine Learning on ESP32-Based Edge Devices

Dalibor Dobrilovic(✉)

Technical Faculty "Mihajlo Pupin", Djure Djakovica Bb, University of Novi Sad, 23000 Zrenjanin, Serbia
`dalibor.dobrilovic@tfzr.rs`

Abstract. Nowadays we are witnesses to the rapid expansion of smart cities and smart transportation services. Smart transportation services especially in the urban area can be different, but they have the same goal, to enhance the driver experiences in the urban area. Smart transportation services have constant growth. This growth is facilitated by the deployment of long-range wireless technologies such as LoRa and LoRaWAN. At the same time, the concepts of Edge computing shape the architecture and services of future smart city applications. This paper describes the implementation methodology of machine learning on the ESP-32 NodeMCU edge devices. The proposed methodology includes the implementation of classifiers such as Random Forrest, Decision Tree, XGBoost, Gradient Boost, and SVC with the scikit-learn library and Python. Scikit-learn-based models are pre-trained on PC and further processed with a micromlgen library to build an Arduino code for implementation on ESP32-based edge devices. The findings of this paper can be implemented in any smart transportation service where the prediction of wireless signal coverage for mobile users is important.

Keywords: edge intelligence · ESP32 development boards · open-source hardware

1 Introduction

In recent decades there has been evident growth of smart cities and smart transportation technology. The advances in the area allow the development of various services that can significantly increase the driver's experience by optimizing the parking process, increasing driver safety, and preventing road accidents. The characteristics of these services are that they are deployed in the urban environment and that they require client mobility. Considering the requirement for wide coverage in urban areas as well as client mobility the utilization of LP-WAN technologies such as LoRa or NB-IoT is needed.

This paper targets the issue of the importance of Machine Learning (ML) on edge devices, integrated into a smart city environment such as smart transportation, smart parking, or similar services that require user mobility and support LoRa or LoRaWAN technology. The goal of this paper is to propose the methodology for the implementation of machine learning on ESP-32-based edge devices. The proposed methodology uses the scikit-learn library and Python for implementing classifiers such as Random Forrest, Decision Tree, XGBoost, Gradient Boost, and SVC. Scikit-learn-based models are pre-trained on PC and further processed with a micromlgen library to build an Arduino code for implementation on ESP32-based edge devices.

This paper presents the example of using machine learning (ML) with Edge Devices capable of predicting the signal coverage, according to the current mobile client position (coordinates), the distance between the mobile client and the base node, and the mobile LoRa receiver speed which is integrated into mobile vehicles. This solution can be used for various mobile and vehicle-based smart city services in any application where it is required to predict the possibility of signal loss or signal coverage. This paper uses positive experience in using Arduino and ESP8266-based platforms in Industrial IoT (Dobrilovic et al., 2021), implementation of ML regressors on ESP8266-based platforms (Dobrilovic et al., 2024), and classifiers (Dobrilovic, 2023).

2 Related Work

The related work for this research covers a broad range of areas, such as the deployment of ESP32-based development boards in IoT systems, implementation of machine learning (ML) on ESP32-based and clone development boards, and edge intelligence.

The paper (Oliveira et al., 2024) gives a review of the emerging concept of the Internet of Intelligent Things (IoIT) covering different areas interesting for this paper, such as embedded systems, machine learning, and edge computing. The paper gives a good overview of existing development boards and their evolution. It also describes the ESP8266, and compares it with previous MCUs, outlining its advance in processing power and memory capacity. This board was designed with an integrated Wi-Fi interface (IEEE 802.11b/g/n), which offers the perspective and the possibility of deploying these boards as a part of the Internet and Wi-Fi networks. The numerous benefits introduced with next-generation ESP32 boards are dual-core CPU, higher frequencies and support for Bluetooth, and Bluetooth Low Energy besides existing Wi-Fi. The Wi-Fi/BLE/BT support enhanced the integration of these boards in the Internet of Things (IoT) and Wireless Sensor Networks (WSN).

The paper (Azevedo et al., 2023) gives an example of implementing a TinyML-based approach. This solution is used for face mask detection and processing of still images. Face mask detection is performed on images of people passing through a monitored area acquired with a low-resolution camera in cooperation with a motion sensor. The goal of this research is to decentralize the decision-making and move it to the edge of the system. The goal of this approach is to improve the overall performance of face mask detection systems. This solution uses pre-trained MobileNet ML architecture.

The article (Gamazo-Real et al., 2023) describes the implementation of ML models from a perspective of resource usage. This research used MLP-ANN in combination with low-cost sensors and commercial off-the-shelf (COTS) processing devices. Furthermore, this article evaluated two estimation methods of environmental parameters on different edge-IoT architectures, and it also focuses on energy efficiency. The solution uses Message Queuing Telemetry Transport (MQTT) as a messaging protocol and Message Passing Interface (MPI).

The article (Siddiqui et al., 2023) presents a novel algorithm using ML to provide integration of multiple sources of data and a different view on diseases. The goal of this research is to develop a low-cost, dependable, IoT and ML-based health monitoring system. The system enables the monitoring of the vital signs of patients and the data analyses based on collected data.

The description of the LoRa mesh network is given in (Giménez et al., 2023). The proposed network offers supporting services at the level of network and application layers. The federated learning application is deployed on tiny network nodes. This paper proposes smart retrainable embedded compute nodes interconnected with a LoRa mesh network creating tiny edge-distributed applications. The authors conducted analyses of the machine learning application performance with the embedded ML model that is trained on the microcontroller board with the federated learning technique. The research uses the experimental environment with three Arduino Portenta H7 boards.

The research (Sneineh et al., 2023) proposes a novel decision-making system for farmers. The proposed system performs measurements to provide data for the farmers. Farmers can use this data to monitor environmental parameters, manage their facilities remotely, and evaluate the effectiveness of the hydroponic system for preventing crop loss.

The manuscript (Márquez-Vera et al., 2023) has in focus the justification of using ESP32 microcontroller in a teaching process. This tool is used for teaching signal processing, control, and data collection. The paper also outlines its communication capabilities making it an applicable solution for educational purposes. With Arduino IDE, as a development tool, the teaching of the usage of sensors and actuators as well as microcontroller programming is available for students with affordable platforms. The findings of this study can be used as a model for possible integration of the implementation of ML to the edge devices in the teaching process.

3 Methodology

This methodology as a development platform uses the ESP32-based microcontroller boards. Its efficiency and usability in this research are justified by the previous examples. Furthermore, since the appliance of this methodology targets the mobile transportation IoT application in the urban environment, the methodology assumes the usage of the LoRa technology and therefore LoRa modules integrated with the ESP32 systems.

3.1 The ESP32-Based Microcontroller Boards

The ESP32-based development boards have ESP32-WROOM-32 module (Espressif Systems, 2018). This is a microcontroller unit (MCU) with Wi-Fi 802.11b/g/n, Bluetooth, and Bluetooth Low-Energy (v4.2 BR/EDR and BLE) connectivity. With its connectivity features the board is suitable for deployment in a wide variety of networked applications. These appliances can even cover high-demand low-power sensor networks with support for signal processing such as voice encoding, music streaming, etc. The module uses an ESP32-D0WDQ6 chip with two CPU cores. The chip supports various energy-saving modes (e.g. modem-sleep, light-sleep, deep-sleep), to rationalize the module power consumption. Therefore, it can be used in various WSN and IoT applications with long lifecycle requirements and in battery-powered applications. ESP32 has an SD card interface and integrated capacitive touch, hall, and temperature sensors. Also, it has several serial buses: SPI, UART, I2S, and I2C. The pinouts of this board are shown in Fig. 1. The figure is partially generated with the usage of Fritzing, a very useful tool for Arduino and clone-based prototyping for educational and other purposes.

The support for Bluetooth (BT), Bluetooth Low Energy (BLE), and IEEE 802.11b/g/n Wi-Fi connectivity allows its appliance for diverse applications. In combination with the processing power, and energy-efficient working modes this module can be functional in a variety of prototypes and even real systems. These systems because of BT, BLE, and Wi-Fi support can vary from short-range to middle-range wireless networks. Because the module has the possibility of connecting the other communication modules via serial buses it can be used even for long-range communication, e.g. for LoRa networks. The big advantage of the module is energy efficiency because in the sleep mode ESP32 chip uses less than 5 μA, making it suitable for battery-powered and wearable electronics applications. The maximum data rate of the ESP32 is 150 Mbps, and the output power is 20.5 dBm.

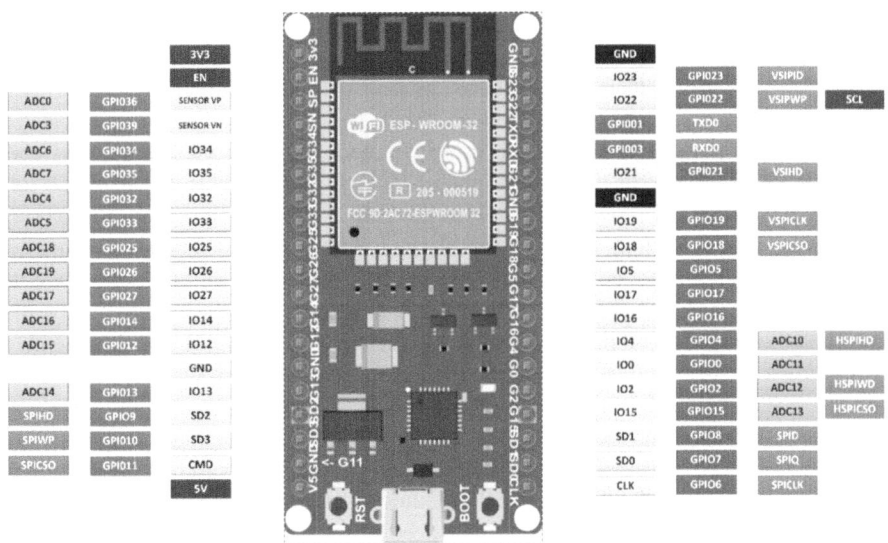

Fig. 1. ESP32-based 38-pin development board and its pinouts.

The ESP32 peripherals are (Fig. 1.):

- 18 Analog-to-Digital Converter (ADC) channels - ADC1 (8 channels, GPIOs 32 - 39), and ADC2 (10 channels, GPIOs 0, 2, 4, 12 – 15, and 25 - 27)
- 3 x SPI – e.g. SPI0 or HSPI (MOSI/GPIO13, MISO/GPIO12, SCLK/GPIO14, CS/GPIO15)
- 3 x UART

- 2 x I2C channels - SDA (GPIO 21); SCL (GPIO 22)
- 2 x I2S channels – e.g. I2S0 output can be set to the DAC's output channels (GPIO 25 & GPIO 26)
- 16 x PWM channels (all pins except GPIO 34 – 39)
- 18 × 12-bit Analog-to-Digital Converter (ADC) channels – ADC1 (GPIO 36, 37, 38, 39, 32, 33, 34. 35); ADC2 (GPIO 4, 0, 2, 15, 13, 12, 14, 27, 25, 26)
- 2 x 8-bit Digital-to-Analog Converters (DAC) channels - DAC1 (GPIO25), DAC2 (GPIO26)
- 10 Capacitive sensing GPIO ports (GPIO 4, 0, 2, 15, 13, 12, 14, 27, 33, 32)

The ESP32 module has a 12-bit ADC (Analogue to Digital Converter) giving the possibility to register the analog sensor output in the range from 0 to 4,905, Together with 2x12-bit SARs it offers 18 analog channels (Espressif Systems, 2020).

3.2 The System Description

This research focuses on the implementation of ML methods on mobile edge devices to predict the signal strength of wireless nodes deployed in the urban environment. The paper uses positive experience in designing and prototyping smart transportation systems in urban environments (Dobrilovic et al., 2021) as well as experience in the utilization of LoRa and LoRaWAN technology (Dobrilovic et al., 2021). There are numerous examples of other papers on urban smart transportation systems such as (Waheb et al., 2024; Devalal et al., 2018; Ortiz et al., 2020; Santana et al., 2023).

Considering the need for urban coverage and mobility of the clients, the LoRa/LoRaWAN as one of the most popular LP-WAN technologies is considered for this system (LoRaWAN 2017, LoRa and LoRaWAN, 2019). This research aims to use the Received Signal Strength Indicator (RSSI) data, as well as the position of the mobile client in the city and its speed to predict whether the signal from a certain LoRa node in the city network will be received. The hypothetical smart urban transportation network is presented in Fig. 2. With the deployment and the coverage of LoRa base stations.

The implementation of ML on mobile clients in a given network scenario is possible with the usage of ESP32-based development boards. These boards are appropriate for utilization because of their low cost, life cycle and physical capacities, computational power, and connectivity capabilities. Together with the ESP32 board, the GPS module should be connected to the board. A GPS module is needed for acquiring data on the GPS position of the mobile client, and the speed of the client. Optionally, the ESP32 board can be equipped with an accelerometer.

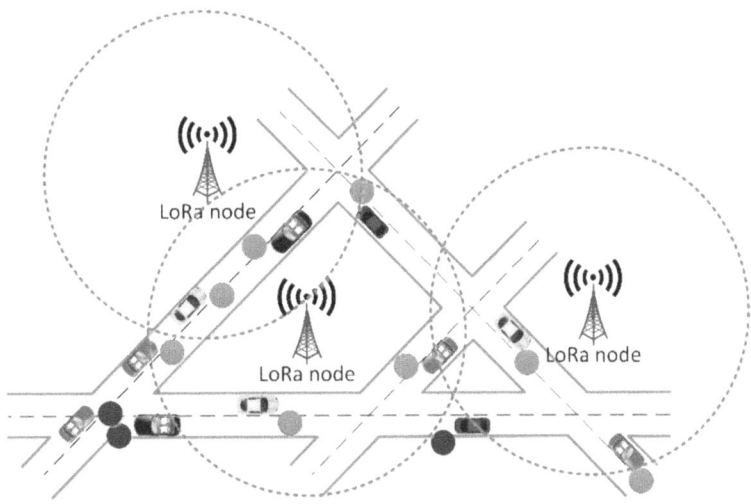

Fig. 2. The hypothetical LoRa-based smart urban transportation network.

3.3 Implementation of Machine Learning

The final phase in this methodology is the implementation of Machine Learning (ML) on edge devices, thus achieving the Edge Intelligence (EI) (Cao et al., 2020). The methodology is graphically presented in Fig. 3.

The important component in this methodology is a TinyML framework that supports ML at embedded edge devices adopting the code to their limited processor and memory capacity. A good overview of TinyML frameworks can be found in (Iborra. R.S. et al., 2020; Ray, P.P. 2022). One of these frameworks is micromlgen (MicroML, 2023). This solution uses the following classifiers: Random Forrest, Decision Tree, XGBoost, Gradient Boost, and SVC. The enlisted classifiers are both supported with micromlgen library for porting ML classifiers made with the support scikit-learn library to Arduino.

This research (Fig. 3, step 1) uses a third-party dataset (Kharkevich, U., Minakov, I., 2020) which is used in the CityScan experiment within the Fed4Fire + project. The data set uses LoRa technology data (Devalal S., Karthikeyan, A., 2018), and only one out of three datasets. The dataset with LoRa SF (Spread Factor) is set to 12. It contains 9,490 entries organized in the following columns:

- node name (node),
- node coordinates (node_lat, node_lon),
- timestamp (ts),
- num,
- signal strength (rssi),
- mobile tag id,
- mobile tag coordinates (tag, tag_lat, tag_lon),
- mobile tag speed (tag_speed),
- distance from LoRa gateway (dist_m), and
- receiving signal indication (received – as a Boolean value).

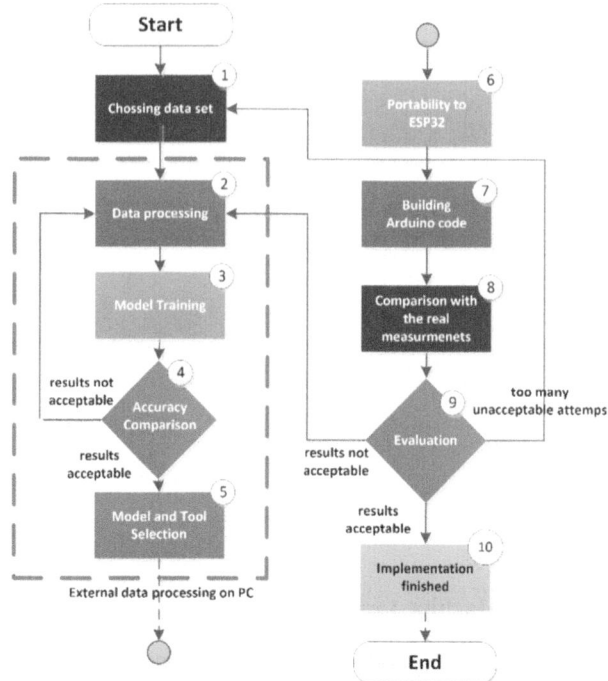

Fig. 3. Methodology for implementing Machine Learning on ESP32-based edge devices.

Scikit-learn, an open-source machine learning library is used in this research for building models for five classifiers (Pedregosa, 2011, Dobrilovic et al., 2023). The dataset is trained to predict receiving signal indication (TRUE or FALSE). The results of classifier accuracy are shown in Table 1 when 30% of dataset rows are used as a test set, and 70% are used as a training set (Fig. 3, steps 2–5).

Table 1. Comparison of five training sets

	RF	DT	XGBoost	Gaussian NB	SVC
Accur	0.773	0.739	0.773	0.705	0.711
F1-score	0.658	0.623	0.657	0.610	0.557

The comparison of five training sets and their confusion matrices for the tested models are shown in Table 2.

The created models for enlisted classifiers are used with the micromlgen library to convert the models to Arduino code (Fig. 3, steps 6–7). The creation of models with such numerous training sets (6.748 rows) in their original form was not successful for the ESP-32 NodeMCU-based platform. As an illustration, the header file for the Random

Table 2. Comparison of five training sets confusion matrix

	RF		DT		XGBoost		Gaussian NB		SVC	
	TRUE	FALSE	TRUE	FALSE	TRUE	FALSE	TRUE	FALSE	TRUE	FALSE
TRUE	622	353	613	362	620	355	658	317	516	459
FALSE	294	1578	380	1492	291	1581	524	1348	363	1509

Forrest classifier is 54 MB, possibly exceeding limitations. This issue will be investigated in the next phases of the project.

Because of that, the next step in the research continues with the discretized training set. Discretization of the dataset is made with the following values:

```
tag_lat:51.217, 51.219, 51.221, 51.223, 51.225
tag_lon: 4.409, 4.411, 4.413, 4.415, 4.417
tag_speed: 1.5, 6, 8, 13, 18, 23, 28
dist_m: 100, 200, 300, 400, 500, 600, 700, 800
```

The accuracy of models with discretized data sets is shown in Table 3. And it is slightly less accurate than in the previous case (Fig. 3, repeated steps 6–7).

Table 3. Accuracy of 70% discretized complete training set

	RF	DT	XGBoost	Gaussian NB	SVC
Accur	0.730	0.730	0.731	0.674	0.708
F1-score	0.619	0.618	0.620	0.612	0.596

The most accurately created models (Random Forest, Decision Tree, and XGBoost) are converted to Arduino code for the ESP32 NodeMCU microcontroller board. This board was chosen because of its better performance compared to other Arduino-like and MCU boards, low prices, availability, and a large supporter community. Because the author experienced some trouble with converting the XGBoost model to Arduino code, only two models (Random Forest and Decision Tree were implemented). The implemented models on the ESP32 board haven't performed well, since they had an accuracy of 56% on 100 test sample measurements (Table 4; Fig. 3, step 8).

Table 4. Confusion matrix of ESP-32 for all nodes implementation

	DT	
	TRUE	FALSE
TRUE	43	44
FALSE	0	13

Because of the detected problems, this research continues with the reduction of data set to only one node (node04), instead of using all 10 nodes in the research (Kharkevich,

U., Minakov, I. 2020). The reduced dataset contains only 949 entries (562 False and 387 TRUE). The accuracy of all five models is shown in Table 5 and the confusion matrices are shown in Table 6 (Fig. 3, step 9). Because the estimation accuracy is raised to about 70%, the decision is to continue the process with this tailored dataset.

Table 5. Accuracy of 70% discretized for Node 04 only dataset

	RF	DT	XGBoost	Gaussian NB	SVC
Accur	0.779	0.779	0.779	0.702	0.611
F1-score	0.732	0.732	0.732	0.632	0.152

Table 6. Confusion matrix of training sets for Node 04 only dataset

	RF		DT		XGBoost		Gaussian NB		SVC	
	TRUE	FALSE	TRUE	FALSE	TRUE	FALSE	TRUE	FALSE	TRUE	FALSE
TRUE	86	31	86	31	86	31	73	44	10	107
FALSE	32	136	32	136	32	136	41	127	4	164

4 Results

As can be noticed in Table 5, the Random Forest, Decision Tree, and XGBoost all have the accuracy of around 78% and the further implementation of ML with micromlgen package is continued with those three classifiers. During the process of porting models to Arduino code (Fig. 3, step 10), because of some problems detected with the XGBoost classifier, only Decision Tree, and Random Forrest classifiers are successfully ported to the ESP32-based microcontroller. The implementation of XGBoost, and a detailed investigation of possible reasons for its unsuccessful porting should be completed in the future.

The next step in the evaluation of the methodology was the testing of Decision Tree and Random Forest implementation. The evaluation is made with the same values for tag GPS coordinates, speed, and distance as in the test dataset. Those values are used as the input values for the ESP32-based microcontroller and its firmware with newly integrated ML models. The result is the prediction of receiving signal indication (with values TRUE or FALSE) meaning whether the signal can be received or not. The results are then compared to real test values for signal coverage. After the comparison of predicted and real values, the results are as follows.

The accuracy of Decision Tree implementation was 98% based on the 100 test entries. The accuracy of Random Forrest is similar. The confusion matrix of the Decision Tree classifier on the ESP32-based platform is shown in Table 7.

The presented results justify the usage of the proposal in the described systems and scenarios. The current limitation of the research findings is that it is based on one LoRa node data. The usability of the proposed method with multiple nodes as data sources should be investigated and analyzed in the future.

Table 7. Confusion matrix of ESP implementation

	DT TRUE	FALSE
TRUE	58	0
FALSE	2	40

5 Conclusion

This paper proposes the methodology for the implementation of machine learning classifiers on edge nodes in sensor networks. The ESP32-based NodeMCU device is chosen as an edge device because of its performance, low cost, and connectivity capabilities. As Python libraries, scikit-learn, and micromlgen are used for porting multiple machine-learning classifiers on an Arduino-like platform. The implementation of Decision Tree and Random Forrest classifiers was successful achieving an accuracy of 98% on a given sample of 100 entries.

These results showed that the implementation of ML can be done on ESP32-based boards with tailored and well-prepared data. The results also proved that the implementation of ML is possible on edge devices based on ESP32 chips, thus achieving Edge Intelligence (EI). Some difficulties faced during this research and addressed in the paper will be further investigated.

The research findings are important because the utilization of ESP32-based boards means that the proposed methodology can be used not only for prototyping real systems but also for educational purposes. The presence of ESP32 and similar boards in the schools especially at higher educational institutions, gives the possibility of using this methodology in the teaching process for ML, AI-supported WSN, IoT, and Edge Intelligence systems.

The current limitation of the research findings concerning the single-node data and methodology expansion to the multiple-node support should be investigated and analyzed in the future.

Acknowledgments. N/A.

Disclosure of Interests. The author declares that he has no competing interests that are relevant to the content of this article.

References

Dobrilovic, D., Brtka, V., Stojanov, Z., Jotanovic, G., Perakovic, D., Jausevac, G.: A model for working environment monitoring in smart manufacturing. Appl. Sci. **11**(6), 2850 (2021). https://doi.org/10.3390/app11062850

Dobrilovic, D., Pekez, J., Ognjenovic, V., Desnica, E.: Analysis of using machine learning techniques for estimating solar panel performance in edge sensor devices. Appl. Sci. **14**(3), 1296 (2024). https://doi.org/10.3390/app14031296

Dobrilovic, D.: Implementing AI on microcontrollers in fog and edge architectures. In: 4th Annual International Conference on Data Science, Machine Learning and Blockchain Technology - AICDMB 2023, 16th-17th March, Mysuru, India (2023)

Oliveira, F., Costa, D.G., Assis, F., Silva, I.: Internet of intelligent things: a convergence of embedded systems, edge computing and machine learning. Internet Things **26**, 101153 (2024). https://doi.org/10.1016/j.iot.2024.101153

Azevedo, M.B., de Medeiros, T. de A., de A. Medeiros, M., Silva, I., Costa, D. G.: Detecting face masks through embedded machine learning algorithms: a transfer learning approach for affordable microcontrollers. Mach. Learn. Appl. **14**(10), 100498 (2023). https://doi.org/10.1016/j.mlwa.2023.100498

Gamazo-Real, J.C., Fernández, R.T., Armas, A.M.: Comparison of edge computing methods in internet of things architectures for efficient estimation of indoor environmental parameters with machine learning. Eng. Appl. Artif. Intell. **126**(Part D), 107149 (2023). https://doi.org/10.1016/j.engappai.2023.107149

Siddiqui, S.A., Ahmad, A., Fatima, N.: IoT-based disease prediction using machine learning. Comput. Electr. Eng. **108**, 108675 (2023). https://doi.org/10.1016/j.compeleceng.2023.108675

Giménez, N.L., Solé, J.M., Freitag, F.: Embedded federated learning over a LoRa mesh network. Pervasive Mob. Comput. **93**, 101819 (2023). https://doi.org/10.1016/j.pmcj.2023.101819

Sneineh, A.A., Shabaneh, A.A.A.: Design of a smart hydroponics monitoring system using an ESP32 microcontroller and the internet of things. MethodsX **11**, 102401 (2023). https://doi.org/10.1016/j.mex.2023.102401

Márquez-Vera, M., Martínez-Quezada, R., Calderón-Suárez, A., Rodríguez, R.M., Ortega-Mendoza, M.A.: Microcontrollers programming for control and automation in undergraduate biotechnology engineering education. Digital Chem. Eng. **9**, 100122 (2023). https://doi.org/10.1016/j.dche.2023.100122

Espressif Systems, ESP32-WROOM-32, Datasheet, Version 2.4, Espressif Inc. A, 2018. https://www.mouser.com/datasheet/2/891/esp-wroom-32_datasheet_en-1223836.pdf. Accessed Feb 2024

Espressif Systems, ESP-IDF Programming Guide, Analog to Digital Converter, Espressif Systems (Shanghai) CO., LTD, 2020. https://docs.espressif.com/projects/esp-idf/en/v4.2/esp32/api-reference/peripherals/adc.html. Accessed Feb 2024

Dobrilović, D., Brtka, V., Jotanović, G., Stojanov, Ž., Jauševac, G., Malić, M.: The urban traffic noise monitoring system based on LoRaWAN technology. Wireless Netw. (2021)

Dobrilović, D., Brtka, V., Jotanović, G., Stojanov, Ž., Jauševac, G., Malić, M.: Architecture of IoT system for smart monitoring and management of traffic noise. In: EAI MMS 2020 - 5th EAI International Conference on Management of Manufacturing Systems, October 27–29, Opatija, Croatia (2020)

Jabbar, W.A., Tiew, L.Y., Ali Shah, N.Y.: Internet of things enabled parking management system using long range wide area network for smart city. Internet Things Cyber Phys. Syst. **4**, 82–98 (2024)

Devalal, S., Karthikeyan, A.: lora technology - an overview. In: 2018 Second International Conference on Electronics, Communication and Aerospace Technology (ICECA), Coimbatore, India, pp. 284–290 (2018)

Ortiz, F.M., de Almeida, T.T., Ferreira, A.E., Costa, L.H.M.K.: Experimental vs. simulation analysis of LoRa for vehicular communications. Comput. Commun. **160**, 299–310 (2020)

Santana, J.R., Sotres, P., Pérez, J., Sánchez, L., Lanza, J., Muñoz, L.: Assessing LoRaWAN radio propagation for smart parking service: an experimental study. Comput. Netw. **235**, 109962 (2023). https://doi.org/10.1016/j.comnet.2023.109962

LoRa Alliance Inc., LoRaWAN® Specification v1.1, 2017. https://lora-alliance.org/resourcehub/lorawanr-specification-v11. Accessed May 2020

Semtech Corporation, LoRa and LoRaWAN: A Technical Overview. https://lora-developers.sem tech.com/library/tech-papers-andguides/ lora-and-lorawan/ Accessed May 2020

Cao K.; Liu Y., Meng, G., Sun Q. An overview on edge computing research. IEEE Access **8**, 85714–85728 (2020)

Iborra, R.S., Skarmeta, A.F.: Tinyml-enabled frugal smart objects: challenges and opportunities. IEEE Circuits Syst. Mag. **20**(3), 4–18 (2020). https://doi.org/10.1109/MCAS.2020.3005467

Ray, P.P.: A review on TinyML: state-of-the-art and prospects. J. King Saud Univ. Comput. Inf. Sci. **34**(4), 1595–1623 (2022)

MicroML homepage, https://www.tinyml.org/about/. 22 Feb 2024)

Kharkevich, U., Minakov, I. (2020). LoRa RSSI vs distance (1.0.0). https://doi.org/10.5281/zen odo.3686907 Accessed Feb 2024

Dobrilovic, D., Bogdan, R., Ognjenovic, V., Marcu, M.: Analyses on usage of MLP regression with WSN data for predicting room occupancy. In: IEEE 19th International Conference on Intelligent Computer Communication and Processing (ICCP), Cluj-Napoca, Romania, pp. 131–136 (2023). https://doi.org/10.1109/ICCP60212.2023.10398671

Pedregosa, F., Varoquaux, G., Gramfort, A., Michel, V., Thirion, B.: Scikit-learn: machine learning in python. J. Mach. Learn. Res. **12**, 2825–2830 (2012)

Stock Open Price Prediction of Software Companies in the BSE SENSEX 50 Index

Chhaya Sonar[1] and Ahmed M. Al Hammadi[1,2](✉)

[1] B. A. M. University, Chhatrapati Samhajinagar, Maharashtra, India
ahmedaqlan593@gmail.com
[2] Mahrah University, Maharah, Yemen

Abstract. The price prediction of the stock is crucial, and it is an equally difficult task due to the extreme fluctuations in the stock market. In this article, the effectiveness of various machine learning algorithms, such as Neural Network, SVM regression algorithm, Random Forest, Regression Tree, KNN regression algorithm, Gradient Boosting Algorithms, Lasso & Ridge regression, and Elastic Net Regression, are employed for predicting stock open prices for software companies listed on the BSE Sensex 50 of Indian stock market. Even though it is not possible to predict open prices very precisely, intrigued by the importance of accurate stock price predictions for financial well-being, the study undertakes a comparative approach to evaluate the performance of these algorithms systematically. This comparison aims to provide insights into algorithms that demonstrate superior performance. The superiority of Ridge Regression and Lasso Regression over other algorithms is observed. The results of this study will be expedient for practitioners and researchers in the finance field. The most influencing factors for predicting stock prices are also obtained.

Keywords: Stock market · SVM Regression Algorithms · Lasso · Ridge regression · BSE index · Random Forest · Regression Tree · KNN

1 Introduction

Predicting stock prices helps investors to make smart choices and ensure smart gains. Using historical data and stock price patterns, trends, and cycles, stock price prediction may help to categorize the top-performing firms in the stock market. Stock price forecasts assist investors in purchasing, selling, or holding stocks for financial benefits. The importance of the stock market is increasing day by day in India. The rate of investors and investments is also growing rapidly over the past few years. The stock market is very volatile, and there are several factors that influence it. Hence, it is very difficult but equally important to map the trends in the market for near-perfect decision-making with the lowest possible level of error while maximizing the net profits. If trends can be predicted accurately using some suitable methods, the returns to the investors can be maximized. Hence, several financial institutes invest their quality efforts in developing mechanisms that would help them make accurate decisions by employing precise predictive techniques to the stock prices. Investors may reduce the risk and maximize rewards with accuracy in the price forecasts. By predicting stock price swings, investors may alter their portfolios, trade timing, and investment strategies to make profits or avoid losses. Stock price forecasting may also assist investors in spotting undervalued or overpriced companies, improving investment option selections [1, 2]. Researchers use various machine learning algorithms in order to evaluate how well several prediction models (such as Neural Network, SVM algorithm regression, Random Forest, Regression Tree, and KNN regression algorithm) can predict the changes in stock prices. While using these methods, a training dataset and a holdout dataset are taken from the whole dataset as part of the data analysis process. Model parameters and specifications are defined in the training dataset and tested and compared in the holdout dataset for use in out-of-sample evaluations [3]. Several regression methods for sales forecasting, including Linear Regression algorithms, Random Forest, KNN, SVR, and Extra Tree Regression, are used apart from emphasizing the significance of ensemble learning models for precise sales forecasting. [4]. Several metrics are used to assess the efficacy of the algorithms, including recall, precision, accuracy, F1 Score, R-squared value, mean squared error, and root mean squared error. To provide more accurate and resilient predictions, ensemble approaches mix several models to improve the accuracy of stock forecasting models by fine-tuning their hyperparameters [5].

2 Objectives of the Study

Researchers attempt to gain deeper insights into stock market behaviour over the time elapsed with effective stock prediction models. This process includes the following objectives: 1. To analyze historical data of the software companies and use it for training and validation purposes. 2. To find which machine learning algorithm models should be used to determine the best algorithm to forecast the open price of software companies. 3. To examine the results and analyze the efficiency of each model evaluation metric. 4. To determine the variables that are important for forecasting open prices.

3 Literature Review

The detection of stock prices is affecting financial investments. This information helps companies to optimize their operations to boost their profits. The results also help investors for equity market decisions [6]. Random forest, LSTM, and neural networks are used as stock price forecasting algorithms. These models are used by [7] to study data from the Shanghai Composite Index with a large number of technical indicators to anticipate stock market index movement. In the study NIFTY 50 Index stock market forecast methods, the researchers examined how feature selection and hyper-parameter optimization affect stock market performance and price predictions to find the best forecasting models, characteristics, and epochs' effectiveness [1]. Deep learning and gradient-boosted decision trees are, among other machine learning methods, to forecast the movement of stock prices and detect financial crises [8]. A boosting method is used in "Boosting the Anatomy of Volatility" to assess the conditional volatility of financial data [9]. By modelling stock market fluctuations using recurrent neural networks and feeding them a continuous data stream, it is possible to construct a proof-of-concept system that can anticipate stock market values [10]. ANNs and SVMs, or support vector machines, are used to forecast the performance of the stock market. The direction of the daily closing price change in the index is used to assess the success of the models [11, 12]. To analyze the data, the authors utilized R^2 to quantify neural network model prediction abilities. Testing of Support Vector Regression (SVR) as a prediction method is used to explore the potential risk-adjusted profits for investors by improving the accuracy of price forecasting [13]. The best model for predicting stock index prices during the COVID-19 era is the autoregressive random forest (AR-RF (1)) model. For the COVID-19 time frame, the Random Forest model successfully forecasts stock index prices [14]. The CART algorithm was used to divide financial stock data into two categories: rising and decreasing stocks. The CART regression tree method is a tree-based machine learning system capable of handling nonlinear correlations in data. The findings revealed that the CART algorithm could efficiently categorize the stock data and determine whether the market trend was growing or declining. This study also used the SVM technique to categorize financial stock data. SVM is a supervised machine learning approach that solves nonlinear problems by translating data into a higher-dimensional space using a kernel function. The findings indicated that the SVM method beat the CART algorithm in terms of classification accuracy as well as error metrics like classification error and mean squared error. The researchers discovered that the SVM algorithm outperformed the CART method [15]. Lasso regression is a penalized least squares estimator approach that uses variable selection and regularisation to enhance the regression model's predictability and interpretability. It works by introducing a penalty term into the residual sum of squares, which is proportional to the absolute value of the regression coefficients. This favours sparse solutions, in which certain coefficients are reduced to zero, so variable selection is conducted. Ridge regression is a penalized least squares estimator technique in which the penalty term is proportional to the square of the regression coefficients. This favours coefficient shrinkage towards zero, but unlike Lasso, it does not provide sparse solutions. Ridge regression is excellent for dealing with multicollinearity, which occurs when the predictor variables are highly correlated. Elastic Net regression combines Lasso and Ridge regression. It contains both the L1 penalty (absolute value of coefficients) from Lasso and the L2 penalty (square of coefficients) from Ridge. This enables Elastic Net to inherit Lasso's variable selection capabilities while also handling multicollinearity, similar to Ridge regression. Elastic Net is very beneficial when there are many more predictors than data [16]. The Elastic Net is a penalized linear regression approach that combines the L1 (LASSO) and L2 (Ridge) penalties under the umbrella of the Elastic Net. In high-dimensional stock market data, where the number of stock market prices (predictors) often exceeds the number of observations, this approach is widely recognized as a strategy for variable selection [17]. The technique known as k-nearest Neighbours (kNN) may be used to make predictions about stock values. The k-nearest neighbours algorithm is a kind of algorithm that can generate predictions by analyzing the k data points that are closest to a particular data point. It is said in the context that the kNN regression approach, which is used to predict numerical values, has been shown to have a higher level of accuracy than the moving average method when it comes to predicting company stock prices [18]. To forecast the IHSG, the researchers employed both the K-NN and ensemble K-NN methods. Gold prices, the rupiah-to-dollar exchange rate, and the Dow Jones Industrial Average (DJIA) index were used as predictor variables. The findings revealed that the ensemble K-NN technique had higher prediction accuracy, with an average forecast of 6078.634 and a Mean Absolute Percentage Error (MAPE) of 7.16%, which is deemed high [19]. Regression using Elastic Nets, Ridges, and Lassoes. These approaches rely on linear regression. Offers helpful information about the weight of various indicators, which helps to comprehend how they impact stock prices. Our comprehension of these dynamics is enhanced by the integration of sophisticated statistical research methods like Lasso with significance analysis [5]. The problem of variable multicollinearity and the process of selecting the most critical variables for study are both handled by these models [20, 21]. Assesses forecasting models using statistical measurements like root mean squared error (RMSE) and R-squared, as well as economic performance indicators like cumulative returns, Sharpe ratios, and utility gains. To enhance the precision of prediction models, statisticians and machine learning experts use the Least Absolute Shrinkage and Selection Operator (LASSO) for feature selection and regularization [22]. When it comes to stock market prediction, Ridge regression may assist in smoothing out the model's predictions by removing fewer essential variables, which might result in more consistent and robust forecasts. [23].

4 Materials and Methods

This research endeavours to forecast the SENSEX 50 stock price of the software companies listed in the Indian stock market. Historical stock market data is needed to build the prediction model. Having a trustworthy source of data that is suitable for forecasting is of the utmost importance, as it provides access to a variety of financial data and analyses currently in modern formats. To make price predictions in the future, www.bseindia.com, this website has all the relevant historical data facts. In addition, it gives the BSE's overall performance as well as the performance of other types of enterprises. For the present study, five years of data of SENSEX 50 index prices from the Bombay Stock Exchange for the period one month 20/11/2023 to 20/12 2023, six months 20/06 /2023 to 20 /12 /2023, one year 01/01/2022 t0 31/12/2022, two years 01/01/2022 to 31/12/2023 and five years 01/01 /2019 to 31/12/ 2023 is considered. The variables under study are the stock's Price, open Price, the highest trading price of the stock during a specific period (high price), the lowest trading price of the stock during a specific period (low price), "PREVCLOSE" What the stock was trading for at the end of the prior trading day, "LTP" that is "last traded price," which is the stock's most recent trading price, "CLOSE PRICE" specifies the stock's price when it closed for the trading day. A stock's "52W L" or "52-Week Low" is its lowest trading price in the last 52 weeks or one year. Like the 52-week high, the 52-week low shows the lowest place the stock has gone over that period, and it's useful as a benchmark. It sheds light on the stock's recent volatility and price range from the past, "VWAP" stands for the volume-weighted average price, which has been determined by considering both the price and volume of trades, "VOLUME" refers to the overall quantity of shares or contracts that were exchanged during a certain time frame, the number of transactions performed is a measure of trading activity. The "OPEN" Price, which is the stock's opening price at the start of each trading day, is the goal variable. The fundamental transaction dataset has seven variables, as mentioned above. Analysts utilize this data to forecast when stock prices will rise or fall. The techniques used for the prediction of open prices are neural networks, support vector machines (SVMs), regression, random forests, regression trees, KNN regression algorithm, gradient boosting algorithms, Lasso and Ridge regression, and elastic net regression. The Lasso Regression & Ridge Regression have the lowest error measures, such as MSE, RMSE and the highest R^2. Also, the dataset was split as 80/20 between training and testing data. The next step was to use data preprocessing methods to get the data ready to input into the models. Next, the model was trained using the selected features. Then, the test data was entered, and the features were measured. Finally, accuracy ratings were produced by feeding the input into several models.

5 Results and Discussion

5.1 Performance Evaluation of Algorithms for Stock Open Price Predictions for Different Time Periods

Different time periods are considered to forecast the open prices of companies to get the best estimates with minimum errors. Hence, the performance of the algorithms is compared by considering data monthly, semi-annually, annually, for two years and for five years.

Regression Evaluation Metrics (REM) is prepared based on machine learning algorithms viz. Neural Network, SVM regression algorithm, Random Forest (RF), Regression Tree (CART), KNN regression algorithm, Gradient Boosting Algorithms (GB), Lasso & Ridge regression, and Elastic Net Regression (EN).

Table 1. Shows comparative results based on one month of data considered from 20/11/2023 to 20/12/2023 with respect to MSE, RMSE, and R^2. It is observed that Lasso Regression has the highest value of R^2 and the lowest value of MSE and RMSE for all companies. The lowest RMSE and greatest R^2 values for Lasso regression indicate better performance than neural networks, support vector machines (SVMs) regression, random forests, regression trees, KNN regression algorithm, gradient boosting algorithms, and elastic net regression. The RMSE values for Lasso and Ridge Regression range between 2 and 28, which are quite small and hence better than all other methods, while the R^2 is greater than 0.85 for all companies.

Table 2. Shows comparative results based on six months of data considered from 20/06/2023 to 20/12/2023 with respect to MSE, RMSE, and R^2. It is observed that Lasso and Ridge Regression have the lowest RMSE and greatest R^2 values, indicating better performance than neural networks, support vector machines (SVMs) regression, random forests, regression trees, KNN regression algorithm, gradient boosting algorithms, and elastic net regression. The RMSE values for Lasso and Ridge Regression range between 1.9 and 11.8, while the R^2 exceeds 0.967.

Table 3. Shows comparative results based on one year of data considered from 01/01/2022 to 31/12/2022 with respect to MSE, RMSE, and R^2. It is observed that Lasso and Ridge Regression have the lowest RMSE and greatest R^2 values among models for each company, indicating better performance than neural networks, support vector machines (SVMs) regression, random forests, regression trees, KNN regression

Stock Open Price Prediction of Software Companies   161

Table 1. Comparision of different algorithms for monthly data

Open Price	REM	Neural Network	SVM	RF	CART	KNN	GB	Ridge	Lasso	EN
HCL TECH	MSE	923247	672	3572.7	2192.5	4716.1	3332.4	61	23	670
	RMSE	960	25.9	59.77	46.82	68.67	57.727	7	4	25
	R^2	−128.9	0.9052	0.497	0.6913	0.3360	0.5308	0.9913	0.9966	0.9057
INFY	MSE	434691	641	1854.7	2516.4	2930.2	1343.3	129	131	583
	RMSE	659.31	25	43.06	50.163	54.131	36.651	11	11	24.16
	R2	−150.6	0.7762	0.3529	0.1220	−0223	0.5313	0.9549	0.9542	0.7963
TCS	MSE	1172636	9352	16914	14139	18548	15594	840	59	3332.6
	RMSE	3424	96.70	130.1	118.90	136.19	124.87	28	7	57
	R^2	−434.7	0.6524	0.3714	0.4745	0.3108	0.4205	0.9688	0.9977	0.8762
TECHM	MSE	330600	213	1138.5	2104.9	1630.6	786.03	308.1	53.110	260.3
	RMSE	574.97	14	33.742	45.879	40.381	28.036	17.6	7.28	6.13
	R^2	−159.1	0.8967	0.4486	-0.0194	0.2103	0.6193	0.8507	0.9742	0.8739
WIPRO	MSE	18965	3.23	127.9	50.738	195.06	83.256	4	5	30.8
	RMSE	137.71	1.797	11.30	7.1231	13.966	9.124	2	2.4	5.4
	R^2	−41.63	0.9927	0.7125	0.8859	0.5614	0.8128	0.9907	0.9869	0.9308

Table 2. Comparison of different algorithms for six monthly data.

Open Price	REM	Neural Network	SVM	RF	CART	KNN	GB	Ridge	Lasso	EN
HCL TECH	MSE	1611.4	81.94	68.02	96.19	341.22	88.62	39.79	48.29	185.64
	RMSE	40.14	9.05	8.25	9.81	18.47	9.41	6.31	6.95	13.62
	R^2	0.7875	0.9892	0.991	0.9873	0.955	0.9883	0.9948	0.9936	0.9755
INFY	MSE	7800.86	74.86	48.02	75.17	197.89	70.64	36.04	48.81	119.19
	RMSE	88.32	8.65	6.93	8.67	14.07	8.4	6	6.99	10.92
	R^2	−2.04	0.9708	0.9813	0.9707	0.9228	0.9724	0.9859	0.9809	0.9535
TCS	MSE	25346.4	354.8	160.06	399.92	1166.22	182.95	125.38	139.52	503.25
	RMSE	159.21	18.84	12.65	19.99	34.15	13.52	11.19	11.8	22.43
	R^2	−0.954	0.9726	0.9877	0.9691	0.91	0.9859	0.9903	0.9892	0.9612
TECHM	MSE	524.12	53.05	82.79	125.26	183.58	61.84	25.93	30.16	138.2
	RMSE	22.89	7.28	9.09	11.19	13.55	7.86	5.09	5.49	11.76
	R^2	0.7883	0.9786	0.9665	0.9494	0.9258	0.975	0.9895	0.9878	0.9442
WIPRO	MSE	82.28	3.35	6.46	5.42	10.71	8.08	3.69	6.44	11.03
	RMSE	9.07	1.83	2.54	2.33	3.27	2.84	1.92	2.54	3.32
	R^2	0.5844	0.9831	0.9673	0.9726	0.9458	0.9592	0.9813	0.9674	0.9443

algorithm, gradient boosting algorithms, and elastic net regression. The RMSE values for Lasso and Ridge Regression range between 3.3 and 71.8, while the R^2 exceeds 0.996.

Table 3. Comparison of different algorithms for yearly data.

Open Price	REM	Neural Network	SVM	RF	CART	KNN	GB	Ridge	Lasso	EN
HCL TECH	MSE	67.83	149.53	204.81	305.01	896.67	181.72	74.5	65.66	670.23
	RMSE	8.23	12.22	14.31	17.46	29.94	13.48	8.6	8.1	25.88
	R^2	0.9989	0.9977	0.9968	0.9953	0.9863	0.9972	0.9988	0.9989	0.9897
INFY	MSE	177.39	183.37	171.12	332.11	567.87	257.63	97.93	111.59	493.83
	RMSE	13.31	13.54	13.08	18.22	23.83	16.05	9.89	10.6	22.22
	R^2	0.9935	0.9933	0.9938	0.9879	0.9793	0.9906	0.9964	0.9959	0.982
TCS	MSE	1148.45	804.02	755.54	1336.33	2508.03	767.07	262.71	318.6	1645.36
	RMSE	33.88	28.35	27.48	36.55	50.08	27.69	16.2	17.8	40.56
	R^2	0.9866	0.9906	0.9912	0.9844	0.9707	0.991	0.9969	0.9963	0.9808
TECHM	MSE	63.75	95.84	119.45	189.14	394.64	130.17	48.02	52.3	257.83
	RMSE	7.98	9.79	10.92	13.75	19.86	11.4	6.92	7.23	16.05
	R^2	0.9955	0.9932	0.9915	0.9866	0.972	0.9908	0.9966	0.9963	0.9817
WIPRO	MSE	54.72	15.64	26.23	35.13	124.42	34.34	11.01	14.11	141.33
	RMSE	7.39	3.95	5.12	5.92	11.15	5.86	3.31	3.75	11.88
	R^2	0.995	0.9986	0.9976	0.9968	0.9887	0.9968	0.999	0.9987	0.9872

Table 4. Comparision of different algorithms for two yearly data

Open Price	REM	Neural Network	SVM	RF	CART	KNN	GB	Ridge	Lasso	EN
HCL TECH	MSE	116.09	49.91	56.38	197.16	164.72	62.44	25.66	32.34	138.66
	RMSE	10.77	7.06	7.51	14.04	12.83	7.9	5.06	5.67	11.77
	R^2	0.9909	0.996	0.9956	0.9845	0.987	0.9951	0.9971	0.9975	0.989
INFY	MSE	210.61	139.66	124.7	175.29	391.79	154.34	90.14	101.13	299.25
	RMSE	14.51	11.81	11.16	13.24	19.79	12.42	9.49	10.1	17.29
	R^2	0.9913	0.9933	0.994	0.9916	0.9812	0.9926	0.9957	0.9951	0.9856
TCS	MSE	1508.9	471.28	465.4	615.49	1293.5	492.49	181.93	249.06	780.78
	RMSE	38.84	21.709	21.57	24.81	35.97	22.19	13.49	15.8	27.942
	R^2	0.9595	0.9873	0.9875	0.9835	0.9653	0.9868	0.979	0.9933	0.979
TECHM	MSE	61.12	113.59	106.8	167.76	590.04	135.47	49.28	52.49	250.34
	RMSE	7.81	10.6	10.33	12.95	24.29	11.64	7.02	7.24	15.822
	R^2	0.9975	0.9953	0.9956	0.9932	0.9759	0.9945	0.9979	0.9979	0.9898
WIPRO	MSE	8.678	6.67	7.11	12.15	41.28	11.99	4.869	5.9628	45.787
	RMSE	2.94	2.58	2.67	3.49	6.42	3.46	2.2	2.44	6.766
	R^2	0.998	0.9985	0.9984	0.9973	0.9909	0.9974	0.9989	0.9986	0.9899

Table 4. Shows comparative results based on two years of data considered from 01–01-2022 to 31–12-2023 with respect to MSE, RMSE, and R^2. It is observed that Lasso and Ridge Regression have the lowest RMSE and greatest R^2 values among models for each company, indicating better performance than neural networks, support vector machines (SVMs), regression, random forests, regression trees, KNN regression algorithm, gradient boosting algorithms, and elastic net regression. The RMSE values for Lasso and Ridge Regression range between 2.2 and 15.8 while the R^2 exceeds 0.993.

Table 5. Comparision of different algorithms for five yearly data

Open Price	REM	Neural Network	SVM	RF	CART	KNN	GB	Ridge	Lasso	EN
HCL TECH	MSE	1190.6	1177.4	1164.5	1304.4	1429.4	1162.2	1111.8	1119.19	1683.21
	RMSE	34.5	34.313	34.13	36.116	37.81	34.091	33.34	33.45	41.0269
	R^2	0.9739	0.9742	0.9744	0.9714	0.9686	0.97448	0.9756	0.9754	0.963
INFY	MSE	67.438	119.403	87.488	141.311	551.366	170.84	81.94	74.25	1278.3
	RMSE	8.212	10.927	9.353	11.887	23.48	13.07	9.052	8.617	35.753
	R^2	0.9995	0.9991	0.9994	0.999	0.9962	0.9988	0.9994	0.9994	0.9911
TCS	MSE	509.3	482.96	393.8	871.8	2768.4	674.8	263.3	261.06	3000.74
	RMSE	22.6	21.97	19.8	29.5	52.62	25.97	16.23	16.16	54.78
	R^2	0.9987	0.9988	0.999	0.9978	0.9929	0.9983	0.9993	0.9993	0.9924
TECHM	MSE	54.498	105.527	65.84	142.866	794.99	156.399	48.626	50.12	741.3
	RMSE	7.382	10.272	8.114	11.952	28.195	12.505	6.973	7.079	27.227
	R^2	0.9994	0.9988	0.9992	0.9984	0.9909	0.9982	0.9994	0.9994	0.9915
WIPRO	MSE	10.353	8.77	8.93	19.548	76.446	29.644	6.717	7.839	160.3
	RMSE	3.217	2.96	2.99	4.421	8.74	5.444	2.59	2.799	12.66
	R^2	0.9994	0.9995	0.9995	0.9989	0.9958	0.9984	0.9996	0.9996	0.9912

Table 5. Shows comparative results based five years data considered from, 01/01/2019 to 31/12/ 2023 with respect to MSE, RMSE, and R^2. It is observed that Lasso and Ridge Regression have the lowest RMSE and greatest R^2 values among models for each company, indicating better performance than neural networks, support vector machines (SVMs), regression, random forests, regression trees, KNN regression algorithm, gradient boosting algorithms, and elastic net regression. The RMSE values for Lasso and Ridge Regression range between 2.6 and 33.5, which are quite smaller and hence better than all other methods, while the R^2 exceeds 0.975.

From all the above comparisons enlisted in Tables 1, 2, 3, 4 and 5, it is observed that the lasso and Ridge regression are the best Algorithms for all data periods for all software companies.

5.2 The Most Influencing Variables for Open Price

It is always of interest and high importance to know which factors influence the open price for the sales and purchase of stocks, as the Lasso regression is the method with low error for prediction of the open price of stocks it is used for further analysis. The most influencing factor(s) among High Price, Low Price, Previous Close, ltp, close Price, VWAP, 52W H, 52 W L, VOLUME, VALUE and number of trades and to get overall better insights, five years of data are considered from 01/01 /2019 to 31/12/ 2023. Lasso regression is used as the variable selection technique for present study. It is a proficient method to find the small subset of features. Lasso regression coefficient graphs are plotted using python represented in Fig. 1.

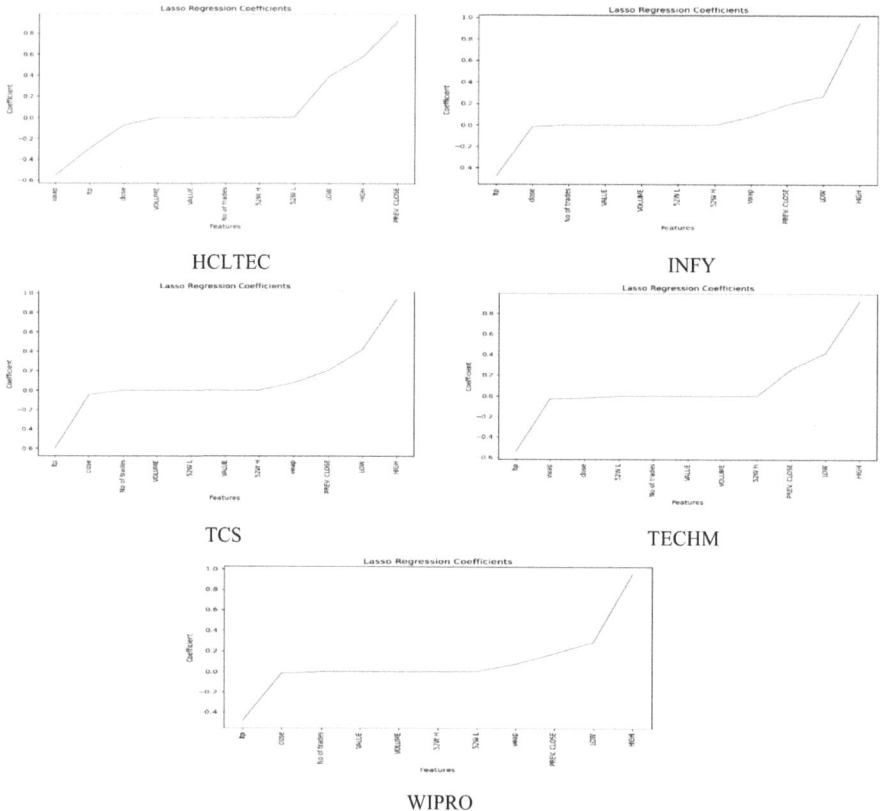

Fig. 1. the relationship between open price and independent variable

From Fig. 1, The relationship between open price is high with the independent variables High Price, Low Price, and previous close for all five software companies.

5.3 The Lasso Regression Prediction Performance for Each Company

Open price prediction performance for Lasso regression is presented in this section. As presented in Fig. 2, The time is shown along the horizontal axis, while the return value is shown along the vertical axis in each subfigure. Whereas the blue colour represents the actual data, the red hue represents the data that was predicted.

From Fig. 2, Due to the fact that the R2 is high and the error rate is low, as was said previously, we were able to comprehend that the prediction performance is great when Lasso regression is used. These observations are related to the five software companies.

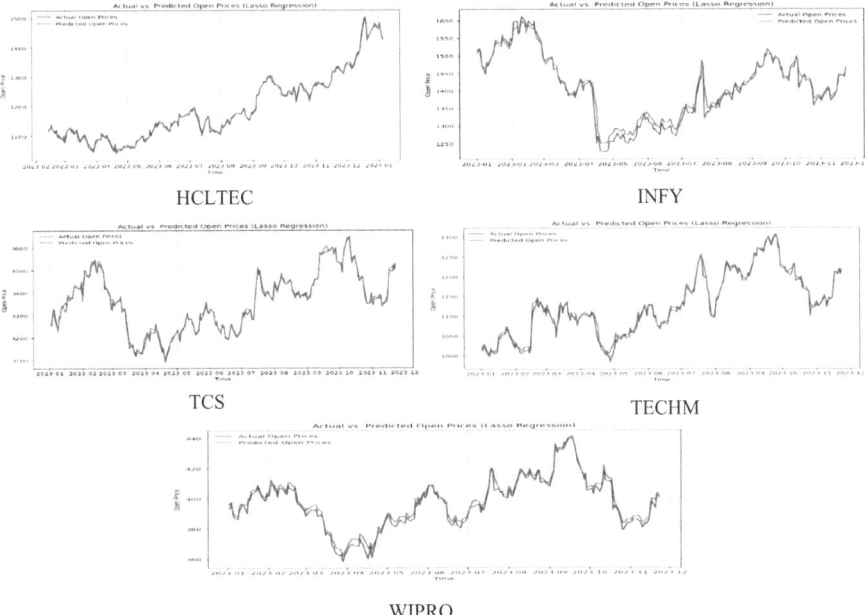

Fig. 2. Actual value vs predicted value for one-year 01/01 /2023 to 31/12/ 2023 in each company using Lasso Regression

5.4 Prediction Five Firms' Stock Open Price Presented Below is a Visual Representation of the Forecast for One year's Worth of Open Pricing.

From Fig. 3, we can observe that the Lasso regression prediction model was almost successful in forecasting the future direction of open prices for the five companies considered. Lasso regression prediction model is best for prediction, also cited in [1, 24, 25].

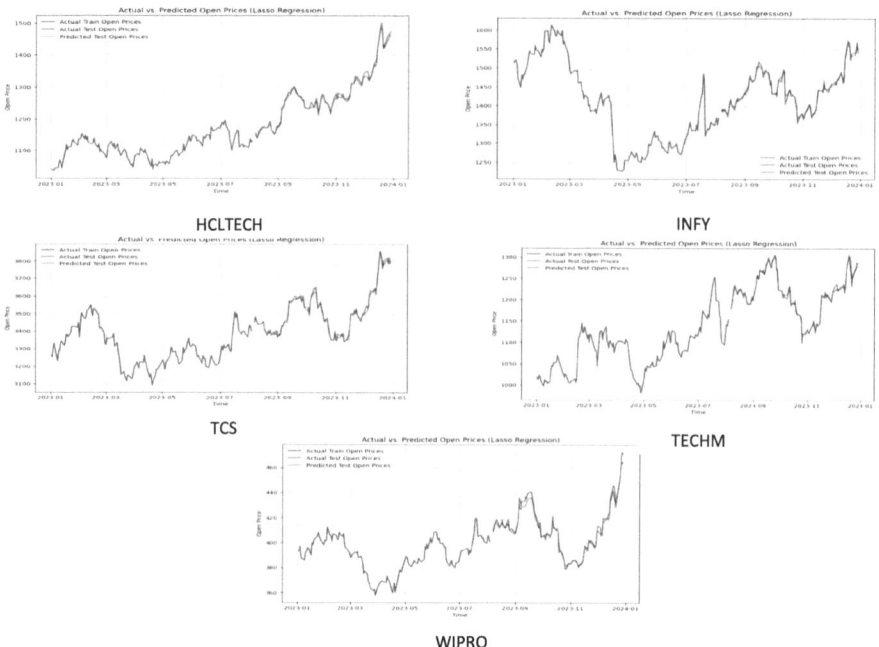

Fig. 3. Prediction of open price for five companies

6 Conclusion

To estimate the open price of the stocks of software businesses that are included in the SENSEX 50 index in the future, Lasso and Ridge Regression techniques are found to be the most effective algorithms. This conclusion is based on the dataset that was used. It achieves the highest R-squared value, which indicates a better fit of the model. The results showed that the Lasso and Ridge model is accurate in forecasting the direction that software firms' stock open prices will go in the future. Other firms that are included in the SENSX50 and other companies that are included in other indices on the BSE might benefit from this outcome. It is also observed that 52W H,52W L, VOLUME, VALUE, and the number of trade Variables are less significant in the group of software companies for estimating the open price. Lasso regression may assist in smoothing out the model's predictions by removing less essential variables, which might result in more consistent and robust forecasts. As a result, The opening price increases when the High Price, Low Price, and previous close increases.

References

1. Fathali, Z., Kodia, Z., Ben Said, L.: Stock market prediction of NIFTY 50 index applying machine learning techniques. Appl. Artif. Intell. **36**(1) 2022. https://doi.org/10.1080/08839514.2022.2111134
2. Ghorbaniid, M., Chongid, E.K.P.: Stock price prediction using principal components. **15**(3) 2020 https://doi.org/10.1371/journal.pone.0230124

3. Huang, C.S., Liu, Y.S.: International journal of economics and financial issues machine learning on stock price movement forecast: the sample of the Taiwan stock exchange. Int. J. Econ. Financ. Issues **9**(2), 189–201 (2019) https://doi.org/10.32479/ijefi.7560
4. Raizada, S., Saini, J.R.: Comparative analysis of supervised machine learning techniques for sales forecasting. Int. J. Adv. Comput. Sci. Appl. **12**(11), 102–110 (2021). https://doi.org/10.14569/IJACSA.2021.0121112
5. Sonkavde, G., et al.: Forecasting stock market prices using machine learning and deep learning models: a systematic review, performance analysis and discussion of implications. Int. J. Financ. Stud. **11**(3) (2023). Accessed 09 Dec 2023. Available: https://doi.org/10.3390/ijfs11030094
6. Cui, Y., Wang, Y., Xu, B.: Analysis of influence mechanism of company stock price based on LASSO-CNN neural network. In: 2022 7th International Conference Intelligent Computing Signal Processing ICSP 2022, vol. 22, pp. 1448–1453 (2022). https://doi.org/10.1109/ICSP54964.2022.9778397
7. Ma, Y.: Stock prediction based on random forest and LSTM neural network (2019). https://doi.org/10.23919/ICCAS47443.2019.8971687
8. Benhamou, E., Jacques Ohana, J., Saltiel, D., Guez, B.: Detecting crisis event with Gradient Boosting Decision Trees (2021). Accessed 12 Dec 2023. [Online]. Available: https://hal.science/hal-03320297
9. Mittnik, S., Robinzonov, N., Spindler, M.: Boosting the anatomy of volatility (2012) Accessed 12 Dec 2023. [Online]. Available: http://www.stat.uni-muenchen.de
10. Kovacs, A., Bogdandy, B., Toth, Z.: Predict stock market prices with recurrent neural networks using NASDAQ data stream. Institute of Electrical and Electronics Engineers Inc., pp. 449–454 (2021). https://doi.org/10.1109/SACI51354.2021.9465634
11. Kara, Y., Acar Boyacioglu, M., Kaan Baykan, Ö.: Predicting direction of stock price index movement using artificial neural networks and support vector machines: the sample of the Istanbul Stock Exchange. Expert Syst. Appl. (2011) https://doi.org/10.1016/j.eswa.2010.10.027
12. Moghaddam, A.H., Moghaddam, M.H., Esfandyari, M.: Stock market index prediction using artificial neural network. J. Econ. Financ. Adm. Sci. **21**(41), 89–93 (2016). https://doi.org/10.1016/J.JEFAS.2016.07.002
13. Henrique, B.M., Sobreiro, V.A., Kimura, H.: Stock price prediction using support vector regression on daily and up to the minute prices. J. Financ. Data Sci. **4**(3), 183–201 (2018). https://doi.org/10.1016/J.JFDS.2018.04.003
14. Bin Omar, A., Huang, S., Salameh, A.A., Khurram, H., Fareed, M.: Stock market forecasting using the random forest and deep neural network models before and during the COVID-19 period. Front. Environ. Sci. **10**, 917047 (2022). https://doi.org/10.3389/FENVS.2022.917047/BIBTEX
15. Hesham, M., Asmaa, I., Jaber, G.: The use of the regression tree and the support vector machine in the classification of the Iraqi stock exchange for the period 2019–2020. J. Econ. Adm. Sci. **28** (132), 74–87 (2022). Accessed 14 Dec 2023. [Online]. Available: http://jeasiq.uobaghdad.edu.iq
16. Al-Jawarneh, A.S., Ismail, M.T., Awajan, A.M., Alsayed, A.R.M., Tahir Ismail, M.: Improving accuracy models using elastic net regression approach based on empirical mode decomposition (2020). https://doi.org/10.1080/03610918.2020.1728319
17. Andu, Y., Hisyam Lee, M., Yahya Algamal, Z.: Adaptive elastic net with distance correlation on the grouping effect and robust of high dimensional stock market price. Sains Malaysiana **50**(9), 2755–2764 (2021) https://doi.org/10.17576/jsm-2021-5009-21
18. Hansun, S.: LQ45 stock index prediction using k-Nearest neighbors regression. Int. J. Recent Technol. Eng. **3**, 2277–3878 (2019). https://doi.org/10.35940/ijrte.C4663.098319

19. Jusman, M., Nur'eni, N., Handayani, L.: Ensemble K-nearest neighbors method to predict Composite Stock Price Index (CSPI) in Indonesia. J. Mat. Stat. dan Komputasi **18**(3), 423–433 (2022). https://doi.org/10.20956/J.V18I3.19641
20. Ruichao, N.: Analysis of influencing factors of Fiscal revenue in Beijing based on Ridge regression and Lasso regression model. Int. J. New Dev. Eng. Soc. **6**(2), 1–5 (2022) https://doi.org/10.25236/IJNDES.2022.060201
21. Iworiso, J.: Forecasting stock market out-of-sample with regularised regression training techniques. Int. J. Econom. Financ. Manag. **11**(1), 1–12 (2023). https://doi.org/10.12691/ijefm-11-1-1
22. Li, X., Liang, C., Ma, F.: Forecasting stock market volatility with a large number of predictors: new evidence from the MS-MIDAS-LASSO model. Ann. Oper. Res. (2022). https://doi.org/10.1007/s10479-022-04716-1
23. Nti, I.K., Adekoya, A.F., Weyori, B.A.: A comprehensive evaluation of ensemble learning for stock-market prediction. J. Big Data **7**(1) (2020) https://doi.org/10.1186/s40537-020-00299-5
24. Rao Polamuri, S., Srinivas, K., Krishna Mohan, A.: Multi model-based hybrid prediction algorithm (MM-HPA) for stock market prices prediction framework (SMPPF) model nonlinear model genetic algorithm artificial neural network and recurrent neural network. Arab. J. Sci. Eng. **45**(3), 10493–10509 (2020) https://doi.org/10.1007/s13369-020-04782-2
25. Wang, X., Wang, W., Zhang, S.: Stock price return prediction based on multifactorial machine learning approaches. In: Proceedings of the 2022 International Conference on Big-data Blockchain and Economy Management (ICBBEM 2022), vol. 5, p. 324. Springer Nature (2022). Atlantis Press, Dec. 2023, pp. 324–333. https://doi.org/10.2991/978-94-6463-030-5_34

Implementation of Morphological Fractional Order Darwinian Operator for Brain Tumour Localization

Kwabena Ansah[1], Wisdom Benedictus Adevu[1], Joseph Agyapong Mensah[2], and Justice Kwame Appati[1](✉)

[1] Department of Computer Science, University of Ghana, Accra, Ghana
{kansah003,wbadevu}@st.ug.edu.gh, jkappati@ug.edu.gh
[2] Department of Computer Science and Information Systems, Ashesi University, Berekuso, Ghana
jamensah@ashesi.edu.gh

Abstract. Over time, Magnetic Resonance Imaging (MRI) has played a pivotal role in accurately delineating brain cancers, facilitating diagnostic processes for medical practitioners. However, the inherent subjectivity in human interpretation presents challenges, prompting the exploration of machine learning models to augment diagnostic guidance. Despite their advantages, these models can be susceptible to errors depending on their objectives. This study addresses segmentation errors induced by data capture noise by introducing the morphological fractional order Darwinian particle swarm optimization (M-FODPSO) approach. Leveraging classical discrete wavelet transform and principal component analysis, pertinent features are extracted from M-FODPSO outputs, subsequently trained on an ensemble classifier. Performance evaluation demonstrates an impressive accuracy of 97.03%, achieved with an average processing time of 1.7161 s, thereby showcasing enhanced tumor cell characterization compared to the classical FODPSO method. This approach effectively mitigates segmentation errors attributed to data noise, underscoring its potential for refining MRI-based brain cancer diagnosis.

Keywords: Brain Tumor · Morphology · Particle Swarm · Segmentation · Optimization · Medical Imaging

1 Introduction

Brain tumors are a significant cause of increased mortality rates across different age groups [18], arising from abnormal cell growth within the brain and classified as either benign or malignant [1, 11]. Benign tumors alter cell structure and function but do not spread to other brain areas [11], while malignant tumors are highly aggressive and can metastasize if untreated, often requiring surgery [1]. Early tumor diagnosis substantially improves survival rates [12], primarily relying on imaging findings. However, accurate interpretation of scans is crucial, given variations in physician skill levels and visual fatigue [12]. Therefore, there is a growing need for reliable electronic classification systems to aid treatment decisions [8], requiring optimal selection of features and classifiers [10]. Despite numerous proposed approaches, challenges persist in accurately segmenting brain tumors [2]. Classical FDOPSO (Fractional-Order Darwinian Particle Swarm Optimization), while historically effective, fails to address noise issues from brain white matter [18], often necessitating manual intervention for segmentation, hindering full automation and complicating patient communication. Improved segmentation techniques are urgently needed to enhance the accuracy and efficiency of brain tumor diagnosis. The subsequent sections of this study are organized as follows: The literature review provides insights into current research efforts and identifies remaining gaps. The materials and methods section outlines the construction of our proposed method and highlights its distinctiveness. The results and discussion section present significant findings and analyze their implications. Finally, the study concludes with recommendations to guide future research and applications in the field.

2 Related Works

In [6], classifiers underwent evaluation using data from the MICCAI BRATS 2016 challenge, which included 3D MRI volumes. Qualitative and quantitative analyses were based on segmentation ground truth [12]. Thirty volumes were randomly chosen from the dataset for classifier training, while 57 volumes were used for assessment. These volumes were preprocessed to a voxel resolution of 1 mm^3 and made accessible in the Virtual Skeleton Database (VSD) [7]. Furthermore, BraTS 2019 was utilized to validate MRI scans from 286 diagnosed brain tumor patients [13]. In a separate study [11], a framework for early detection and classification of Glioblastoma was introduced. Multiple modality MRI images, including those with artifacts, underwent processing using a contrast-stretching method to enhance resolution and edge information, aiming to reduce manual evaluation time for tumor cell assessment. Moreover, [16] stressed the importance of image pre-processing before segmentation and feature extraction, removing extraneous signs and labels from images to avoid adverse effects on the classification process. Other investigations involved CT and MR images in DICOM and JPEG formats from 60 patients diagnosed with GBM. While MR images were typically noise-free, researchers introduced Gaussian and Rician Noise to assess their proposed model's response to noise. Additionally, a registration process aligned MR images to CT images iteratively using an affine method until minimal cost was achieved. Furthermore, in [7], a brain mask was generated using FLAIR volume, followed by intensity normalization and smoothing using a Gaussian filter to standardize voxel properties.

In [15], a method was proposed for efficient feature extraction using convolutional neural networks (CNNs) to identify essential components within images as compact feature vectors. This methodology streamlines tasks such as image matching and retrieval, especially in large-scale image datasets. The approach involved convolving small-sized filters with input patterns to extract distinctive features for network training. Pre-trained architectures were utilized for transfer learning (TL) to capture visually salient characteristics, followed by softmax classification of these features. However, [13] introduced the ACU-NET framework, which enhances feature extraction through depthwise separable convolution modules and dense residual blocks. Experimental results showcased the framework's exceptional segmentation accuracy for brain tumor images, though its adaptability was noted to be somewhat limited. In contrast, the NASNet architecture, integrating convolutional and recurrent neural networks (CRNNs) as an update function, exhibited superior performance over time compared to other architectures like ResNet50 and DenseNet201, using a gradient policy. Challenges arose during training due to insufficient annotated medical image datasets, particularly with deep learning algorithms. To tackle this issue, data augmentation techniques such as flipping and rotation were employed to expand the training dataset. Additionally, the Figshare dataset was utilized for multi-classification of tumors to determine tumor types.

Additionally, in [14], the Stationary Wavelet Transform (SWT) and Principal Component Analysis (PCA) were integrated to develop an image fusion algorithm. This research also combined soft tissue information from MRI with CT scans. Accurate segmentation of medical images was vital for subsequent therapy planning, accomplished through the use of the Enhanced Fuzzy C Means (EFCM) algorithm to segment the fused MRI scans. While PCA reduced database dimensions by identifying significant factors, it lacked translation invariance. To mitigate this limitation, SWT was utilized, modifying filters with zero padding to ensure translation invariance.

3 Methodology

This section lays the groundwork necessary for understanding the research methodologies utilized in crafting a multimodal algorithm for brain tumor detection. It covers essential elements including data acquisition, the proposed M-FODPSO pipeline, feature extraction, and performance evaluation metrics.

3.1 Dataset

In this study, secondary data obtained from KAGGLE [9] was utilized for model evaluation. The images were formatted in JPEG and anonymized for public use. A total of 253 MRI scans were utilized in the study, classified into three categories: benign, malignant, and no tumor.

3.2 System Design

The system design for this study comprises five phases: data acquisition, pre-processing, segmentation, feature extraction and selection, and classification, as illustrated in Fig. 1.

3.3 Image Preprocessing

The study emphasized by [17] highlights the pivotal role of an efficient preprocessing strategy in improving the segmentation of medical images. Acknowledging this necessity, the study initially converted MRI images to grayscale, ensuring uniformity by adjusting gray levels within the range of 0 to 255. Additionally, it's widely recognized that MRI images are prone to noise despite technological advancements. To tackle this issue, the study employed anisotropic diffusion filters (ADF), selected for their ability

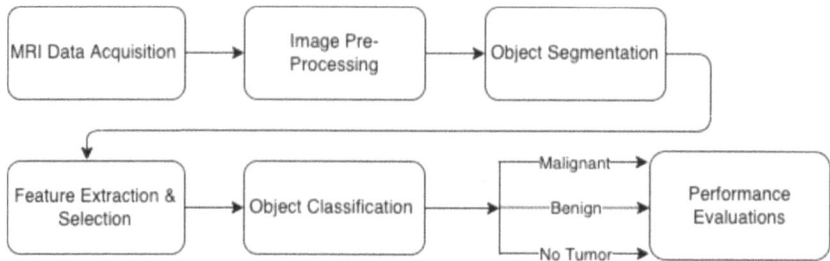

Fig. 1. System Design

to reduce noise while retaining crucial image features such as edges and lines essential for accurate interpretation. Conceptually, ADF functions by treating image edges akin to large contiguous volumes with higher capacitance, requiring more time to equalize temperature compared to smaller volumes or noise pixels. Consequently, ADF effectively removes high-frequency noise while preserving the sharpness of MRI images. The mathematical formulation of ADF is provided by Eq. 1.

$$I_s^{t+1} \approx I_s^t + \tfrac{\lambda}{|\eta s|} \sum_{p \in \eta s} g\left(\nabla I_{s,p}^t, \gamma^t\right) \nabla I_{s,p}^t \qquad (1)$$

After the image enhancement with ADF, the MRI image is resized to 256 × 256 after which skull scraping is performed all as part of image pre-processing.

3.4 Tumor Segmentation

Traditionally, the goal of image segmentation is to divide an image into meaningful regions, with the extent of partitioning determined by the specific requirements of the application [10]. Despite extensive research into various segmentation techniques and algorithms, determining the algorithm that produces the most accurate segmentations remains challenging, whether for individual images, a group of images, or an entire class of images. In this study, a novel approach utilizing morphological fractional-order Darwinian particle swarm optimization (FODPSO) was introduced for image segmentation. This method capitalizes on the benefits of fractional-order Darwinian particle swarm optimization, offering promising potential for enhanced segmentation accuracy. Further insights into this approach are elaborated in Sect. 3.4.1.

3.4.1 Morphological FOD-PSO

The Particle Swarm Optimization (PSO) algorithm draws inspiration from collective behaviors observed in natural phenomena such as fish schooling and bird flocking, embodying the concept of swarm intelligence [3]. As a population-based optimization technique, PSO exhibits versatility across a wide range of applications owing to its ability to address diverse problem sets [3]. It simulates the behavior of animal societies where no single individual serves as a leader within the swarm, yet collective efforts lead to the discovery of optimal solutions [3]. To address the inherent limitations of traditional

PSO algorithms, researchers introduced the Darwinian Particle Swarm Optimization (D-PSO) to refine the natural selection process within PSO frameworks [20]. D-PSO introduces a novel approach by integrating natural selection mechanisms to inherit PSO attributes, thereby improving algorithm performance and mitigating the risk of converging to local optima [20]. Unlike conventional methodologies, D-PSO coordinates multiple parallel PSO algorithms simultaneously, each equipped with a randomized swarm, enabling diverse exploration of solution spaces [20]. Furthermore, to tackle convergence challenges encountered in D-PSO, researchers introduced the Fractional-order DPSO (FOD-PSO) in 2012 [4]. This variant enhances convergence properties by redefining functions associated with inertia weight and acceleration, introducing new topologies, and incorporating innovative optimization techniques into the PSO framework [4]. Through the establishment of new equations governing particle position and velocity, FOD-PSO significantly improves the performance of D-PSO, offering enhanced convergence characteristics [4]. Equations 2 and 3 delineate the mathematical formulations governing particle position and velocity vectors within the framework of FOD-PSO, respectively.

$$D[V(t)] = \frac{1}{T^\alpha} \sum_{k=0}^{r} \frac{(-1)^k \Gamma(\alpha+1) V(t-kT)}{\Gamma(k+1)\Gamma(\alpha-k+1)} \quad (2)$$

$$V_i(t+1) = \alpha V_i(t) + \frac{1}{2}\alpha V_i(t-1) + \frac{1}{6}(1-\alpha)V_i(t-2)$$
$$+ \frac{1}{24}\alpha(1-\alpha)(2-\alpha)V_i(t-3) + c_i r_i (P_i - X_i(t))$$
$$+ c_2 r_2 (P_g - X_i(t)) \quad (3)$$

where α is the fractional order, $X_i(t)$ is the position, r, the particle length, $v(t)$ is the velocity, c_1, c_2 are the weight controls, k is the summation index, P is the particle and t is the time

Algorithmic implementation for the FOD-PSO for image segmentation is as outline as follows:

Step 1: Initialize the parameters of $\alpha, c_1, c_2, x(t)$ and $v(t)$. The particle velocity $v(t)$ was set to zero (0) and the position $x(t)$ was set in random positions and α, c_1, c_2 are set as weights that controls the initial influence of the swarm.

Step 2: Increase the iteration counts and these changes in position and velocity will change into numbers iteration (i) by setting $i = 1,2,3, \ldots,$ n and reaches the best position and velocity.

Step 3: Delete a particle swarm. The particle will be removed if it false below the minimum bound $SC_c(N_{kill}) = SC_{kill}^{max}\left[1 - \frac{1}{N_{kill}+1}\right]$. Once it is deleted, there is a reset on the counter value that approaches the threshold number where in case here is that sum of particles removed from the particle swarms.

Step 4: Spawn a new swarm. It is generated as $p = \frac{f}{NS}$, where f is generated from a random number between 0 and 1 and the NS is the number of swarms.

Step 5: Calculate the velocity and the position. The inertial weight is noted as w and was used to update the velocity vector and position as shown in Eq. 4 and Eq. 5.

$$v_{n+1}^i = wv_n^i + c_1 r_1 \left(g_n^i - x_n^i\right) + c_2 r_2 \left(l_n^i - x_n^i\right) + c_3 r_3 \left(m_n^i - x_n^i\right) \quad (4)$$

$$x_{n+1}^i = x_n^i + v_{n+1}^i \tag{5}$$

Step 6: Update the position of *Pbest* and *Gbest*. It was updated for each particle as in PSO, *Gbest* is global best and *Pbest* is personal best. In order to obtain a best value, Eq. 6 and Eq. 7 are used for *Pbest* and *Gbest*.

$$p_i(t+1) = \begin{cases} p_i(t) \, iff(x_i(t+1)) \geq f(p_i(t)) \\ x_i(t+1) \, iff(x_i(t+1)) < f(p_i(t)) \end{cases} \tag{6}$$

$$Gbest = min\{f(p_0(t)), f(p_1(t)), f(p_2(t)), \ldots, f(p_m(t))\} \tag{7}$$

Step 7: Update new velocity and position is calculated using the equations in step 5.
Step 8: Acquire fitness function. It was used to evaluate how successful to obtain the objectives of the study. Equation 8 is used to maximize the fitness function of the study

$$\varphi^c = max_{1 < t_1^c < \ldots < t_{n-1}^c < L} \sigma_B^{2c}(t^c) \tag{8}$$

Extra pixels in the segmented image after step 8 are removed by morphological erosion using two radius disks shaped structure elements after segmentation to separate weakly connected of the segmented tumor image. To improve the lost pixels, dilation with two structural elements is used with a threshold value of 60.

Step 9: Erosion: Process. The erosion process decreases the picture by removing the backdrop of the foreground pixels. The threshold picture $(S)_c$ is turned into an deleted reconstruction image using the structural component S. The destroyed image is achieved when Eq. 9 is applied.

$$I \ominus S = \{c | (S)_c \subseteq I\} \tag{9}$$

Step 10: Dilation Process: The dilatation function is carried out by thickening the MRI's object of detection (i.e., the tumor component), plugging holes in a structural element, and lastly collecting all of the mined material. The produced images are acquired after the applying Eq. 10

$$I \oplus S = \{c | c = i + s, i \in I, \, s \in S\} \tag{10}$$

3.5 Feature Extraction and Selection

After the tumor segmentation phase, the subsequent step involves extracting and selecting features from the resulting image. Focusing on both the quality of image features and computational efficiency, this study utilizes the discrete wavelet transform (DWT) as the primary method for feature extraction. Following this, the processed image undergoes analysis using the Gray Level Co-occurrence Matrix (GLCM) along with associated texture feature computations. From the GLCM, various shape characteristics of the segmented tumor are derived, including circularity, irregularity, area, perimeter, and shape index, each representing distinct shape attributes. Additionally, intensity characteristics such as mean, variance, standard deviation, median intensity, skewness, and kurtosis are extracted. Furthermore, texture characteristics such as contrast, correlation, entropy, sum of square variance, homogeneity, entropy, and cluster shading are incorporated into the extracted features. Mathematically represented, all these features are defined by equations ranging from Eq. 11 to Eq. 19, providing a comprehensive framework for characterizing the tumor's shape, intensity, and texture attributes.

$$Mean(M) = \frac{1}{m*n} \sum_{x=0}^{m-1} \sum_{y=0}^{n-1} f(x, y) \qquad (11)$$

The mean is the average of the numbers used to describe the brightness of an image. Bright images have the highest mean value, while dark images have the lowest mean value, calculated by adding all the pixel values of the segmented tumor and dividing by the total number of pixels in the image. Standard deviation, on the other hand, represent the probability distribution of a population and can be used to measure inhomogeneity.

$$SD(\sigma) = \sqrt{\left(\frac{1}{m*n}\right) \sum_{x=0}^{m-1} \sum_{y=0}^{n-1} (f(x, y) - M)^2} \qquad (12)$$

On the contrary, entropy is defined as Eq. 13.

$$E = - \sum_{x=0}^{m-1} \sum_{y=0}^{n-1} f(x, y) \log_2 f(x, y) \qquad (13)$$

Kurtosis describes the shape of a random variable's probability distribution, characterized by a parameter and defined as:

$$K_{urt}(X) = \left(\frac{1}{m*n}\right) \frac{\sum (f(x,y) - M)^4|}{SD^4} \qquad (14)$$

Skewness measures the symmetry or lack thereof in the intensity level distribution, indicating the asymmetry about the mean defined in Eq. 15.

$$S_k(X) = \left(\frac{1}{m*n}\right) \frac{\sum (f(x,y) - M)^3|}{SD^3} \qquad (15)$$

Energy quantifies the degree of repetition of pixel pairs within an image. Defined by Haralick's GLCM feature, it is also known as angular second moment and is defined as:

$$En = \sqrt{\sum_{x=0}^{m-1} \sum_{y=0}^{n-1} f^2(x, y)} \qquad (16)$$

The contrast and the inverse difference moment is defined in Eq. 17 and 18 respectively.

$$C_{on} = \sum_{x=0}^{m-1} \sum_{y=0}^{n-1} (x - y)^2 f(x, y) \qquad (17)$$

$$IDM = \sum_{x=0}^{m-1} \sum_{y=0}^{n-1} \frac{1}{1+(x-y)^2} f(x, y) \qquad (18)$$

Correlation assesses the degree of linearity within the tumor image, particularly revealing linear structures such as striations, and is defined as:

$$C_{orr} = \frac{\sum_{x=0}^{m-1} \sum_{y=0}^{n-1}((x,y)f(x,y)) - M_x M_y}{\sigma_x \sigma_y} \quad (19)$$

Consequently, three types of features are extracted, detailing the structural aspects of intensity, shape, and texture. However, these features may exhibit redundancy to mitigate the need for feature reduction.

3.5.1 Feature Reduction

Integrating additional features into image classification tasks can significantly escalate computational complexity and memory storage demands. To tackle this challenge, Principal Component Analysis (PCA) emerges as a valuable technique for reducing dimensionality in datasets with numerous interrelated variables. PCA aims to condense the dataset while retaining as much variance as possible. As one of the most commonly used linear dimensionality reduction methods, PCA scrutinizes the original dataset to identify a combination of input features that effectively summarizes its distribution, thereby reducing its original dimensions. This is achieved through maximizing variances and minimizing reconstruction error, observed by analyzing pairwise distances within the dataset. By adeptly capturing the essential information present in the original data, PCA streamlines computation and storage, thus augmenting the complexity of image classification tasks.

3.6 Classification

Image classification is the process of categorizing images based on their visual content, representing a crucial task within computer vision systems. As described by [19], spectral information embedded in digital integers across one or more spectral bands plays a fundamental role in image classification. This spectral data enables the classification of individual pixels, known as spectral pattern recognition, with the aim of assigning each pixel to specific classes. After feature extraction and selection from MRI tumor images, a total of 12 features derived from Gray Level Co-occurrence Matrix (GLCM) features were selected for classification using trained ensemble classifiers, distinguishing between benign and malignant tumors. Ensemble classifiers are a prevalent approach in machine learning, enhancing performance by aggregating predictions from multiple classifiers. Unlike traditional models, ensemble classifiers combine a variety of classifiers into a unified algorithm or unit. To address the limitations of individual machine learning models, ensemble classifiers utilize various classifiers, including boosted trees, bagged trees, and rusboosted trees, to mitigate error rates. For instance, rusboosted trees integrate AdaBoost with random undersampling to address data imbalance, while bagged and boosted trees employ bagging and boosting strategies, respectively, to enhance decision tree classifier accuracy. These strategies offer expedited training, reduced complexity, and heightened flexibility, making them particularly suitable for large datasets.

3.7 Performance Measure

The performance was evaluated in two aspects: accuracy and sensitivity, calculated using the following formulas:

True Positive (TP): An abnormal brain MRI image correctly identified as abnormal.
True Negative (TN): A normal brain MRI image correctly identified as normal.
False Positive (FP): A normal brain MRI image incorrectly identified as abnormal.
False Negative (FN): An abnormal brain MRI image incorrectly identified as normal.
Accuracy.
Four steps were considered as follows
Step 1: Acquiring of pixel of ROI area
Step 2: Compute Percentage Relative Error

$$PRE(\%) = \frac{|F-E|}{E} * 100 \tag{20}$$

Step 3: Compute Average Percentage Relative Error

$$Average\ PRE(\%) = \frac{\sum_{i=1}^{n} PRE}{n} \tag{21}$$

Step 4: Evaluate accuracy

$$Accuracy(\%) = 100 - Average\ PRE(\%) \tag{22}$$

In terms of sensitivity performance, recognition statistics and the range of relative error were utilized, as displayed in Table 1. Equation 23 yields a sensitivity rate of 91.70%. The table illustrates that more than three-quarters of the images exhibit favorable results, indicating the efficacy of the proposed method in segmenting MRI images.

$$Sensitivity(\%) = \frac{TP}{TP + FN} * 100\% \tag{23}$$

Table 1. Outcome of the recognition statistics

Range of relative error	Recognition statistics	Number of images	TP/FN
(0, 5)	Very good	147	TP
(6, 10)	Good	56	TP
(11, 15)	Average	29	TP
(16, 20)	Below average	10	FN
Greater than 20	Poor	11	FN

4 Results and Discussion

In this section, Table 2 presents the results of five randomly selected segmented images of the original image alongside those processed using the propossed method from the Kaggle dataset.

The transition from grayscale to binary images can be observed through the process of erosion and dilation using the mathematical morphology algorithm. The enhanced FODPSO image effectively segments the image by removing noise, as depicted in 1(b),

Table 2. Output of the proposed method

No	Original image (a)	FODPSO method (b)	M-FODPSO method (c)
1			
2			
3			
4			
5			

2(b), 3(b), 4(b), and 5(b) from the FODPSO method. Additionally, it successfully segments clustered images, as shown in 3(a), and blurry images, as shown in 2(a), into the desired Region of Interest (ROI), demonstrated in 1(c), 2(c), 3(c), 4(c), and 5(c) respectively. These image outputs validate the success of the improved FODPSO method, thus achieving the objective.

Table 3. Assigning parameter values to the proposed FODPSO

Parameter	Optimal Value
α	0.6
c_1	0.8
c_2	0.8
w	1
Number of Swarm	4

Table 3 defined parameters of the FODPSO method. The parameters, adopted from literature, were defined due to FODPSO's utilization of a fractional calculus approach to handle particle convergence with memory effect. This is crucial as the convergence algorithm relies on the fractional order, allowing for optimization of results by controlling parameter values.

4.1 Performance

The performance evaluation consisted of two components: an accuracy of 97.03% and a sensitivity of 91.70%. Among the 253 sampled datasets, it was noted that 73 MRI images contained benign tumors, 153 MRI images contained malignant tumors, and 27 MRI images exhibited no tumors. Sample results can be observed in Fig. 2 and Fig. 3.

Statistics of selected Feature extraction results is shown in the Table 4.

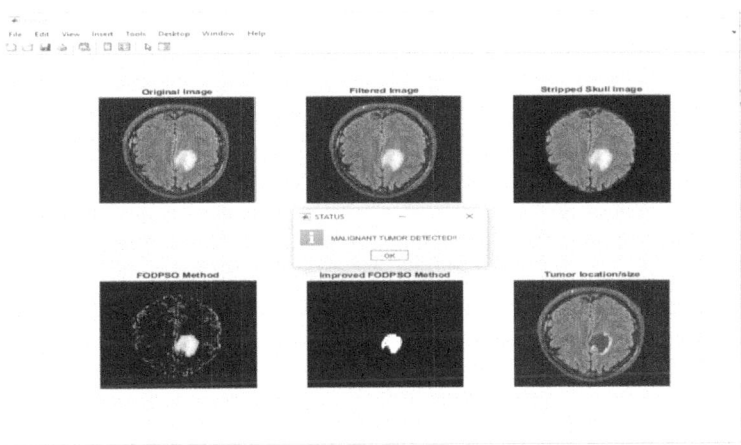

Fig. 2. Sampled tumor detected with Malignant infection

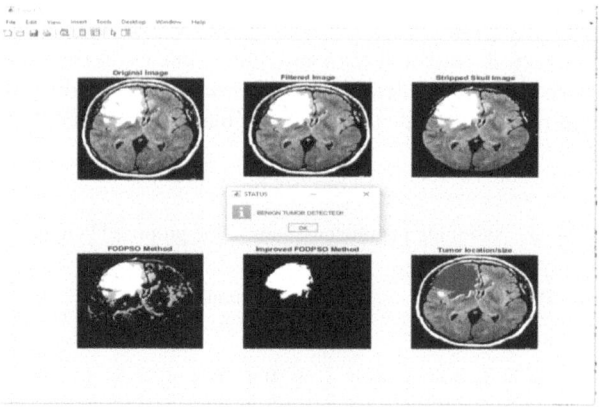

Fig. 3. Sampled MRI image with Benign infection

Table 4. Statistics result of the selected features

No	Contrast	Correlation	Energy	Homogeneity
1	0.4490	0.0754	0.9175	0.9741
2	0.3852	0.0990	0.8992	0.9691
3	0.3913	0.0972	0.8854	0.9663
4	0.4041	0.1118	0.8999	0.9689
5	0.3072	0.1971	0.8883	0.9663

5 Conclusion

Computer vision and image processing techniques are pivotal in diagnosing brain disorders, particularly brain cancers, utilizing high-resolution MRI images. However, manual segmentation of brain tumors proves to be time-consuming. This study focuses on accurately diagnosing early-stage brain tumors to enable timely treatment, given their life-threatening nature. The integration of FODPSO approaches, mathematical morphology, and ensemble classifiers has emerged as a widely adopted method for segmenting and classifying brain tumors from MRI images, enhancing accuracy and effectiveness. This technique combines ADF for image enhancement, FODPSO for segmentation, mathematical morphology for noise removal, and ensemble classifiers for classification, resulting in robust performance.

The proposed method effectively segments brain tumor MRI images by distinguishing background tissues and addressing noise issues introduced during image enhancement. By leveraging mathematical morphology to refine segmentation and remove noise artifacts, the updated FODPSO model demonstrates superior performance compared to manual algorithms, achieving a classification accuracy of 97.03% and a sensitivity of 91.70%. Additionally, the enhanced FODPSO model showcases performance on par with top image categorization models, indicating its potential as a benchmark for algorithm development and highlighting its broader applicability beyond brain tumor classification.

References

1. Alhassan, A.M., Zainon, W.M.: BAT algorithm with fuzzy C-ordered means (BAFCOM) clustering segmentation and enhanced capsule networks (ECN) for brain cancer MRI images classification. IEEE Access **8**, 201741–201751 (2020)
2. Appati, J.K., Owusu, E., Soli, M.A.T., Adu-Manu, K.S.: A novel convolutional Atangana-Baleanu fractional derivative mask for medical image edge analysis. J. Exp. Theoret. Artif. Intell. 1–23 (2022)
3. Atia, N., et al.: Particle swarm optimization and two-way fixed-effects analysis of variance for efficient brain tumor segmentation. Cancers **14**, 1–32 (2022)
4. Couceiro, T.: Introducing the fractional-order Darwinian PSO. Spring-Verlag London Limited **8**(5), 234–239 (2012). https://doi.org/10.1007/s11760-012-0316-2
5. Di Ianni, M., Airan, R.D.: Deep–fUS: a deep learning platform for functional ultrasound imaging of the brain using sparse data. IEEE Trans. Med. Imaging **41**, 1813–1825 (2022)
6. El-Melegy, M.T., El-Magd, K.M.: A multiple classifiers system for automatic multimodal brain tumor segmentation. In: ICENCO 2019 - 2019 15th International Computer Engineering Conference: Utilizing Machine Intelligence for a Better World, pp. 110–114 (2019)
7. El-Melegy, M.T., El-Magd, K.M., El-Baz, A.S.: Adaptive window for automatic classification-based segmentation of multimodal brain tumor. In: 2018 IEEE International Symposium on Signal Processing and Information Technology, ISSPIT 2018, pp. 108–113 (2019)
8. Gumaei, A., Hassan, M.M., Hassan, M.R., Alelaiwi, A., Fortino, G.: A hybrid feature extraction method with regularized extreme learning machine for brain tumor classification. IEEE Access **7**, 36266–36273 (2019)
9. Kaggle: Retrieved from Kaggle (2020). https://www.kaggle.com/datasets/navoneel/brain-mri-images-for-brain-tumor-detection
10. Latha, R.S., Sreekanth, G.R., Akash, P., Dinesh, B., kumar, S.D.: Brain tumor classification using SVM and KNN models for smote based MRI images. J. Crit. Rev. **7**(12), 1–4 (2020)
11. Leena, C., Sreedevi, A.: Framework for multimodal image fusion for detection of glioblastoma. In: 2020 IEEE 17th India Council International Conference (INDICON), pp. 1–6 (2020)
12. Li, M., Kuang, L., Xu, S., Sha, Z.: Brain tumor detection based on multimodal information fusion and convolutional neural network. IEEE Access **7**, 180134–180146 (2019)
13. Ling, T., Wenjie, M., Jingming, X., Sajib, S.: Multimodal magnetic resonance image brain tumor segmentation based on ACU-net network. IEEE Access **9**, 14608–14618 (2021). https://doi.org/10.1109/ACCESS.2021.3052514
14. Nandeesh, M.D., Meenakshi, M.: Tumor detection using enhanced FCM for multimodal brain images. In: 2019 2nd International Conference on Intelligent Computing, Instrumentation and Control Technologies, ICICICT 2019, pp. 1419–1422 (2019)
15. Mehrotra, R., Ansari, M.A., Agarwal, A., Anand, R.S.: A transfer learning approach for AI-based classification of brain tumors. Mach. Learn. Appl. **2**, 100003 (2020)
16. Shantta, K., Basir, O.: Brain tumor detection and segmentation: a survey. IRA Int. J. Technol. Eng. **10**(4), 55–61 (2020). (ISSN 2455–4480)
17. Tariq, S., et al.: Brain tumor detection and multi-classification using advanced deep learning techniques. Microsc. Res. Tech. 1–13 (2021)
18. Thangarajan, S.K., Chokkalingam, A.: Integration of optimized neural network and convolutional neural network for automated brain tumor detection. Sens. Rev. **41** 16–34 (2021)

19. Tharangini, S., Krishna, G.R.: Skin cancer detection using particle swarm optimization. Int. J. Creative Res. Thoughts **6**(2), 2–8 (2018)
20. Tillett, J., Rao, T., Sahin, F., Rao, R.: Darwinian particle swarm optimization. In: Proceedings of the 2nd Indian International Conference on Artificial Intelligence, vol. 1, pp. 1474–1487 (2005)

Author Index

A
Abraham, Anil   123
Adevu, Wisdom Benedictus   169
Akshaya, M.   16
Al Hammadi, Ahmed M.   156
Alkalbani, Abdullah Said   97
Al-Najjar, Tariq   56
Ansah, Kwabena   169
Appati, Justice Kwame   169

B
Bouni, Mohamed   80
Brown, Dane   123

C
Ceh-Varela, Edgar   3

D
Dobrilovic, Dalibor   144
Dornberger, Rolf   68
Douzi, Khadija   80
Douzi, Samira   80

F
Fadzli, Fazliaty Edora   109

H
Hanne, Thomas   68
Hssina, Badr   80
Husain, Hamid   42

I
Imhmed, Essa   3
Ismail, Ajune Wanis   109

J
Javed, Hira   42
Joy, Emmanuel   16

K
Kishore, Amit   27

M
Mensah, Joseph Agyapong   169

O
Om Prakash, C.   27

S
Satyanarayana, Degala   97
Schmid, Yannick   68
Shahid, M Kaab Bin   42
Sharma, Madan Kumar   97
Sonar, Chhaya   156

W
Wahsha, Mohammad   56
Wahsheh, Heider   56

SPRINGER NATURE

GPSR Compliance

The European Union's (EU) General Product Safety Regulation (GPSR) is a set of rules that requires consumer products to be safe and our obligations to ensure this.

If you have any concerns about our products, you can contact us on ProductSafety@springernature.com

In case Publisher is established outside the EU, the EU authorized representative is:

Springer Nature Customer Service Center GmbH
Europaplatz 3
69115 Heidelberg, Germany

The manufacturer's authorised representative in the EU is Springer Nature Customer Service Centre GmbH, Europaplatz 3, 69115 Heidelberg, Germany. If you have any concerns regarding our products, please contact ProductSafety@springernature.com

Printed and bound by CPI Group (UK) Ltd, Croydon, CR0 4YY

26/03/2026

02078933-0011